AN INQUIRER'S GUIDE TO ETHICS IN AI

An Inquirer's Guide to
Ethics in AI

Matthew S.W. Silk
and Ian J. MacDonald

b

broadview press

BROADVIEW PRESS
Peterborough, Ontario, Canada

Founded in 1985, Broadview Press is a fully independent academic publishing house owned by approximately twenty-five shareholders—almost all of whom are either Broadview employees or Broadview authors. Broadview is supported by a collaboration with Trent University, a liberal arts university located in Peterborough, Ontario—the city where Broadview was founded and continues to operate. Broadview is committed to environmentally responsible publishing and fair business practices.

© 2025 Matthew S.W. Silk and Ian J. MacDonald

Library and Archives Canada Cataloguing in Publication

Title: An inquirer's guide to ethics in AI / Matthew S.W. Silk and Ian J. MacDonald.
Other titles: Inquirer's guide to ethics in artificial intelligence | Ethics in AI
Names: Silk, Matthew S. W., author. | MacDonald, Ian J., author.
Description: Includes bibliographical references and index.
Identifiers: Canadiana (print) 20240481690 | Canadiana (ebook) 20240481704 | ISBN 9781554816408 (softcover) | ISBN 9781770489615 (PDF) | ISBN 9781460408858 (EPUB)
Subjects: LCSH: Artificial intelligence—Moral and ethical aspects.
Classification: LCC Q334.7 .S55 2024 | DDC 174/.90063—dc23

Broadview Press handles its own distribution in Canada and the United States:
PO Box 1243, Peterborough, Ontario K9J 7H5, Canada
555 Riverwalk Parkway, Tonawanda, NY 14150, USA
Tel: (705) 482–5915
email: customerservice@broadviewpress.com

Canada**I✦I** Broadview Press acknowledges the financial support of the Government of Canada for our publishing activities.

Edited by Robert M. Martin
Book Design by Em Dash Design

Broadview Press® is the registered trademark of Broadview Press Inc.

PRINTED IN CANADA

Dedicated to Adam Silk and to
Samantha St. Amand. She promised
she'd be there with me when
I paint my masterpiece.

Contents

Acknowledgements

Grateful acknowledgements are due to Samantha St. Amand for her support in this project and for her assistance in creating some of the diagrams. I would also like to acknowledge the assistance of the students of PHIL 228 and peers at the University of Waterloo for their insights and feedback as I developed the material for this book. I'd also like to thank my parents Glenn and Judy for their support.

—Matt Silk

I thank my co-author, Matt Silk, for proposing this project and inviting me to join him in writing this book. The project has been a source of enjoyment and has enriched my understanding of the ethical issues surrounding artificial intelligence. Working on this book together has made me a better thinker and writer. It has also solidified our friendship, which I very much value. I also appreciate the advice and help from Shannon Dea, who provided wise suggestions in the early phases of composing the book and has been a stable source of encouragement.

—Ian MacDonald

We are grateful to Broadview Press for all their support throughout the writing process. Thank you to Leslie Dema from Broadview, who initially listened to our proposal for the book and passed the word along. Also, at Broadview, we greatly appreciate the help of Stephen Latta and

Archie Fields, who handled the drafts, review process, and manuscript and provided sound editorial advice, and of the anonymous reviewers and their feedback.

—Matt & Ian

Introduction

Ethical conversations are of little use if only a tiny fraction of the population understand them. This book is designed so that people from any background can understand the concepts and issues involved with ethics and artificial intelligence. If you come from a computer science or engineering background, the text is organized to provide you with basic tools of inquiry that can be applied to ethics. This text aims to provide insights into what prompts ethical thinking and the means to stimulate critical thinking about real-world ethical issues without getting lost in abstractions. If you come from an ethics or non-technical background, you will find technical accounts of artificial intelligence and statistical modeling that we hope can be easily understood. Our intention is to bridge the gap between scientific and philosophical frameworks and concepts so that the central issues can be understood by anyone.

Our goal is to provide a practical guide to understanding real-world ethical issues involving artificial intelligence. Speculative questions such as whether artificial intelligence can have consciousness, the future of work, and the future of warfare are not addressed. Instead, we offer a toolbox of questions and considerations to assist in the inquiry of ethical issues involving AI by understanding the issues in terms of the value judgements, incentives, assumptions, and logistics that prompt ethical concern about the use of AI.

This text has two central aims. The first is to present an account of inquiry that will be of use to anyone who wishes to explore ethical issues involving AI. The second is to explore the relationship between scientific research into AI and the rest of society. Given that science in general and AI specifically have the potential to significantly affect society for better or for worse, what ethical responsibilities do researchers and developers have to society and to what extent should society regulate the uses of science to avoid or resolve social and moral problems? The first chapter begins with an account of ethical thinking and ethical inquiry, providing an inquirer's toolbox to assist the reader in critically analyzing ethical issues. From there, each chapter will be treated as separate inquiries investigating different topics.

The second chapter inquires into the moral responsibilities of scientists. This will be significant for providing an understanding of the relationship between science, AI development, and society, and for considering the moral responsibilities involved in AI development. These themes will resonate throughout the remainder of the book. In the third chapter we inquire into modelling and machine learning, investigate the meaning of fairness, and discover how AI can produce biased results. In the fourth chapter we will inquire into machine learning and opacity and consider the ethics of belief in terms of whether we ought to believe the output of an AI is accurate. In the fifth chapter we will inquire into the relationship between social ethics, democracy, scientific research, and the effect of AI on democratic norms. The final chapter will consist of a series of inquiries into various issues involving the relationship between humans and AI, including an ethical examination of privacy and how AI can affect human capabilities.

The text will explore AI's impact on industry, commerce, law, medicine, and education, among other sectors. Along the way, as we learn new lessons, we will add additional tools to our toolbox and consider potential practical solutions to moral problems.

Nevertheless, the text is not designed to be the final word on these topics. AI is developing at a rapid rate and the ethical issues involved will no doubt continue to change. Our intention is to offer you the tools of inquiry and an informational background to not only make you knowledgeable about AI-related issues, but, more importantly, to make you more critically minded about them. Although we simply could not discuss every ethically salient issue involving AI, this text will hopefully serve as a resource you can use to make inquiries on these issues on your own. However, the central themes of this text will continue to be relevant because they emphasize the reasoning and valuing processes that underlie artificial intelligence development. You will also find additional online resources to assist you as you conduct your own inquiries related to ethics and artificial intelligence.

Ethical Inquiry

The development of **artificial intelligence** (AI) presents significant ethical challenges. Some are more abstract and hypothetical in nature, such as whether an AI could ever become sentient, and if it did, whether it should have rights. But such problems are more remote because the technology doesn't exist yet. If we look at AI being developed and used today, however, we find a wide spectrum of specific kinds of ethical concerns. Some of these problems are new and unique, such as whether the use of an AI-powered voice assistant has the potential to inhibit social development. Other problems, as we will see, are not new but present a new challenge because of the increasing scale of use.

It may seem like making ethical AI should be a simple matter of embedding ethical principles into the design language. Isaac Asimov's three laws of robotics, which prohibit robots from harming humans, comes to mind.[1] We simply need to develop AI that makes similar ethical decisions. But before we reach that stage, we need to ask ourselves how *you* make ethical decisions? How do you know when you are facing an

---

1   Asimov presents these three laws in his 1942 short story "Runaround":
    1. A robot may not injure a human being or, through inaction, allow a human being to come to harm.
    2. A robot must obey the orders given it by human beings except where such orders would conflict with the First Law.
    3. A robot must protect its own existence as long as such protection does not conflict with the First or Second Law.

ethical choice? What would you think about to help you decide what to do? Imagine that you promise your cousin that you will help them move on a certain date, only to remember that you had already planned to go on vacation with a friend on the same day. How would you decide what is more important? Does it matter that your cousin is your relative? Should you go with the person you made the commitment to first? Who needs your presence more? Even a relatively simple decision can be very complex ethically speaking, so before we consider how to make AI ethical, we need to think more about what ethics is in the first place.

1. Why Engage in Ethical Thinking?

A common question when beginning the study of ethics is to ask why we should be ethical at all. There are at least two important ways we can understand this question. The first is as a practical question; why should I or anyone else bother to be ethical in our actions? The second is as a skeptical one; is there something substantive to ethics other than subjective opinion such that I have an objective reason to act ethically if I weren't otherwise being compelled to act this way? Is their force just a matter of the subjective opinions of those in power? We will briefly consider each question in turn.

For most people the question "why be ethical?" is simply not a practical one. We may not all always act ethically, but since we live near other people, we must pay attention to societal ethical norms or customs. We need others to cooperate with us and we wish to avoid punishment. If you want to live in a society, you must concern yourself sometimes with how you should act, and so almost everyone theoretically must occasionally "act ethically."

But to some this may seem problematic. Shouldn't our intentions matter? If you avoid assaulting someone you don't like simply because you don't want to be punished or shamed, can you really be said to be acting ethically? It really depends how important you think intentions matter by themselves and whether intention should be the sole determination of what counts as ethical. Should we praise the reckless person who ends up harming people they meant to help just because they had good intentions? In the context of AI, where a machine makes decisions, it may not even make sense to speak of intentions anyway.

Focussing too much on intention can be problematic. Where do we draw the line on intention? If I act in ways my community prescribes even though I don't want to simply because I want to function better in my community, am I acting selfishly or selflessly? Does intention have to spring from a pure disposition to be ethical? How would you know if it did? No one can climb inside your head to know what your intentions might be. The problem with focussing on intention is that it often isn't provable one way or the other. Later, we will consider some answers to these questions, but the point is that focussing only on intention can lead us to some confusing and unhelpful thinking. If we put these questions aside, there is still the practical situation to consider.

Another practical reason to behave ethically can perhaps best be described by philosopher William James: "The only force of appeal to us ... is found in the 'everlasting ruby vaults' of our own human hearts, as they happen to beat responsive and not responsive to [an ethical] claim."[2] In other words, the practical reason to be ethical (I suspect for most of us) is that you care about the other living things around you. While this doesn't settle the question of all possible reasons to be ethical, this answer is sufficient so that we can set the question of being ethical at all aside for now.

But if ethics is often based on how people feel about things, then a more skeptical question arises. Is ethics all just a matter of subjective opinion, and if so, why should anyone feel compelled to act according to the feelings of others? If we consider controversial topics like abortion, for example, it might be tempting to conclude that each side of the issue simply has their own feelings on the topic. However, ethical life always takes place in a context and circumstance that makes such perspectives more complicated than simple feelings. The way we value things emerges from our lived experiences and involves issues like collective cooperation, psychological and cultural idiosyncrasies, resource management, and logistics, which necessarily makes discussions about ethics more complicated than subjective feelings.

For example, have you ever seen a film that you did not personally like, but that you can nevertheless appreciate and understand why it is a good film? How could such a thing be? Is the value of the film subjectively in your head, or can something have objective value despite you

---

2   James 1891, 340.

not personally valuing it? Perhaps the film reminds you of a bad personal experience, or you don't like the cast. Maybe your roommate watched the film on repeat, and you simply can't sit through it anymore. It's not a bad film, but you don't like it. Examples like this remind us that values are more complicated than how any one individual feels about something. Perhaps the film has an objective kind of value because it has a kind of universal appeal, or the film does something nuanced and interesting in a way that previous films haven't. We can recognize this objective novel effect on others, while also understanding our own personal distaste. The reason these values have a kind of objectivity that goes beyond private feelings is because they can have a kind of objective effect on the world that goes beyond any one person's idiosyncratic feelings.

Is it merely a subjective opinion that someone shouldn't drive without a license? Assuming that we generally don't wish to harm people, we have developed a system to make driving safer and thus to avoid such harms. If a study were to demonstrate that driver license programs dramatically reduce such harms, does this not constitute an objective reason to get a license? If so, it is because we are investigating real world interactions relating to issues that we care about. The rules, norms, and prescriptions we arrive at are a product of ethical inquiry that goes beyond subjective feeling and reaches out into the real world, to objective existing conditions in the world, and that observes the consequences of our actions. Sometimes, we even find because of such investigations that we realize that what we originally valued is too problematic when put into practice in relation to other things we value.

None of this means that it would never be ethical to drive a car without a driver's license. Sometimes emergencies happen and you might have to take the wheel. But in that case, we recognize the unusual or idiosyncratic nature of that situation, which might objectively call for actions to be taken that wouldn't be acceptable otherwise. But we can also recognize that such situations do not permit simple generalizations about ethics. Ethics is and always will be a grand experiment in finding what kinds of actions work best in different circumstances. In that regard, ethical inquiry done well will be objective in a way that subjective feelings about ethics will not be. For that reason, ethical prescriptions arising from ethical inquiry are worth listening to in circumstances when solutions to ethical problems have earned an experimental prestige.

2. Making an Ethical Decision

How do you recognize that you are dealing with an ethical situation? Why are some actions considered ethical or unethical while other actions are neither? Imagine that you are an AI developer tasked with developing an **algorithm** to determine car insurance rates. How would you know if something you are doing is ethically problematic? Let's say that the algorithm is built on a dataset containing credit scores and you find that because of this it is biased towards giving better rates to wealthier people and worse rates to poorer people. The insurance company may assume that people with higher credit scores are generally safer people, all things being equal. Thus, a wealthier person with a worse driving record may get a better rate than a poorer person with a better record. Does this fact mean that there is an ethical issue here?

To better understand why some kinds of actions take on ethical importance while others don't, we need to consider ethics in a basic way. Let's say you plan to go away for the weekend on vacation. Is this ethically significant? It doesn't really seem like it. But imagine you find out that your cousin suddenly needs help moving that weekend and no one else can help but you. Now you must make a choice; does it suddenly seem like that choice is ethical? Why? Perhaps it is because you are now faced with two incompatible options: to leave on vacation or to help your cousin. Since you cannot do both yet feel compelled to do both, you are faced with a dilemma because you must determine which action is *worth* pursuing.

When faced with a dilemma you could simply proceed by trying any random thing in a trial-and-error fashion, or maybe you could rely on what is socially customary without thinking about it. This is an unreflective approach to ethics. The other option is to try to gather facts about the dilemma, analyze the situation, and use this analysis to propose a solution before testing it. This is reflective ethics and it involves asking ourselves what justification there is for taking one action rather than another. If we were already convinced about what action is justified, we would have no reason to reflect on it. Thus, reflective ethics occurs in cases where there is an open question about what is right or wrong, and it involves reflective inquiry towards finding a justifiable solution.

2.1 REFLECTIVE MORAL THINKING

In reflective moral thinking, to determine which option has greater moral worth, you need to make a value judgement. Value judgements involve consideration of two elements: ends and means. An **end** is a possible goal or settlement of your dilemma. In our previous example, we are faced with a choice between two ends: going on vacation or helping your cousin move. The **means** are the practical and logistical factors that are required to obtain an end. So, the means used to obtain the end of going on vacation might include the cab ride, the plane ticket, travellers' insurance, and so on, while your means of helping your cousin include possibly a truck and your own body. If an action is considered as an end, you need to consider whether you have the means to obtain it, and sometimes this can affect what you choose to value. If you had less money, for example, you might feel more hesitant about going on vacation in the first place, whereas if you had extra money you might choose to hire movers as means to obtain both of your ends: you go on vacation while your cousin gets the help they need.

We can now understand why the auto insurance example has a moral salience. The AI developer tasked with creating an algorithm to determine insurance rates uses credit scores as a means towards their end goal of creating a reliable insurance rate generator (Figure 1). However, the AI developer doesn't wish to negatively impact anyone by creating a system where people are discriminated against for arbitrary reasons. But the means they've chosen has now created a conflict between these two ends. Now the developer must either choose between one of those two ends or find a new means of satisfying both ends without creating further problems.

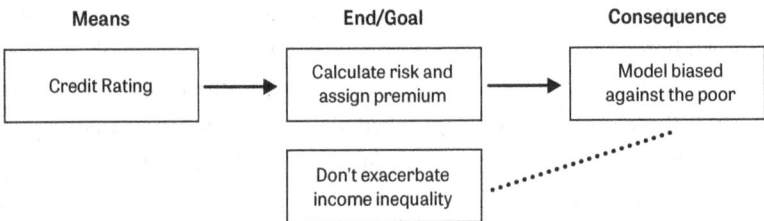

FIGURE 1 · Using credit scores as a means helps bring about a reliable system for calculating insurance premiums, but it produces consequences that conflict with other goals.

This problem may be further compounded if the insurance company commissioning the algorithm insists on using credit scores, because they assume a higher-scoring applicant is less risky. It might also be easier to obtain credit scores compared to the alternative variables. These factors highlight the genuine moral tension that exists in this situation and that drives us to moral reflection and inquiry.

But how do we decide which ends are of greater moral worth? To appreciate the different perspectives, interests, goals, and logistics involved in AI development, we must inquire into situations and scenarios that call for our ethical attention. Thus, the aim of this text is to create a resource for moral inquirers to empower anyone to better study, understand, and prescribe reflective solutions to moral problems: an inquirer's toolbox. We've already learned that understanding the relationship between our end goals and the means we use to obtain those goals can help us understand how and why moral dilemmas come to be. We can add the first tool in this toolbox, which is the use of ends-means reasoning to help us evaluate the consequences of one proposed moral solution compared to another. Next, we will consider the potential use of moral principles in helping us choose among different ends.

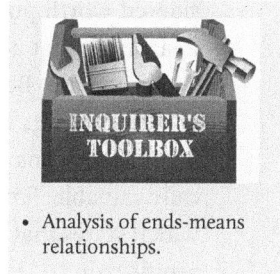

INQUIRER'S TOOLBOX

- Analysis of ends-means relationships.

2.2 MORAL PRINCIPLES

A moral principle is an abstract general rule that can help determine which things are more morally important than other things. They tell us what we *ought* or *should* do. Sometimes these moral principles are general and might apply to anyone, such as "the golden rule" which asks us to treat others as we would prefer to be treated. Moral principles can also be aimed at specific kinds of cases. For example, a professional code of ethics might state that accountants must be honest with their client at all costs. The most well-known ethical principles come from prominent moral theories.

2.2.1 Utilitarianism, Deontology, and Self-Driving Cars

Utilitarianism is a moral theory developed by philosopher Jeremy Bentham (1748–1832). Bentham tried to identify which kinds of ends are always worth pursuing no matter what and then to develop a moral principle that would guide us towards those ends. An end that is considered worth pursuing no matter what is called **intrinsically valuable**. Reasoning that all our actions are either about pursuing happiness and avoiding pain, Bentham articulated the concept of utility, or the property that things have that tend to produce happiness or prevent pain. Since all actions ultimately aim at promoting utility, utility is intrinsically valuable. From this he was able to articulate the **principle of utility**, which states that an action is ethically good if it promotes happiness and avoids pain.[3]

Because there is nothing morally special about your happiness compared to anyone else's happiness, any moral action will need to promote overall happiness rather than just your own. The principle of utility therefore tells us that we ought to act to promote the greatest amount of overall happiness for the greatest number of people. The utilitarian principle might be useful because it tells us what is ethically more important when we have conflicting ends, and this may make it well suited for AI application cases. To consider how the utilitarian moral principle might be applied to a real-world case, we will examine the case of self-driving cars.

The level of automation of a vehicle is commonly denoted using the SAE International's five-level classification of automated driving systems.[4] Most commercially available vehicles that claim to have automated driving are level two, meaning that while the vehicle is capable of acceleration and steering, it requires a human driver to remain attentive and be ready to assert control. A fully autonomous level-five vehicle does not exist, but let's consider a scenario that a future level-five vehicle could face. Imagine that the vehicle must swerve to avoid a car accident on the road and the only available pathway for the vehicle includes a

---

3   Bentham 1781 [2000], 14.
4   (1) Driver assistance for safety or comfort: automatic braking or cruise control. (2) Partial driving automation: switchable additional controls; e.g., speed or lane control. (3) Conditional: driver can switch on full AI control of driving. (4) High automation: as default, driver is passenger, but can switch to manual control when there is any risk. (5) Full automation: AI always drives (but a switch to manual control is provided. See several explanatory websites for SAE International (the Society of Automotive Engineers).

curb with a pedestrian standing on it. If the vehicle tries to stop or continues forward it will hit the scene of the accident where multiple people are trapped inside another vehicle. If it swerves to avoid the accident, it will hit the curb and kill the pedestrian. If you were programming such a vehicle, what would you prefer that it does?

The utilitarian principle suggests we must take the action that produces the greatest amount of utility—"happiness"—to the largest number of people. Assuming each person involved in the situation has an equal capacity to create and feel happiness, then the moral thing to do would be to have the car swerve and kill the pedestrian to save the people on the road since it would save multiple lives at the cost of one. Moral principles like utilitarianism seem helpful when it comes to dealing with ethical issues in AI because notions like "minimize casualties" or "save the largest number of people" seem to provide the morally acceptable answer, and it's simple enough that we could train an AI to follow such instructions. Thus, embedding a moral principle in AI could be a way to ensure that it acts ethically.

But this answer is problematic. Why should some car company get to decide who lives and who dies? Isn't it wrong to sacrifice someone without their consent? To see why, we can consult another well-known moral principle formulated by Immanuel Kant (1724–1804). Kant believed that moral right and wrong should not depend on context; something shouldn't be ethically okay at one time but not another, or okay in one area but not in another. Like Bentham, Kant sought what is intrinsically valuable. To formulate a principle to determine what is universally moral, he reasoned that moral goods are good without qualification or regard to circumstance, so must be discovered by reason alone. The only thing that is good without qualification, Kant reasoned, is our own intention to be good for its own sake. But what does it mean to be good for its own sake? It means having the intention to act morally as a matter of duty rather than for a reward. If giving to the homeless makes you happy, this has no moral status according to Kant unless you are doing it for the sake of moral duty.

Duty-centric ethics are called **deontological**. But what does this moral duty involve? It's based on reason alone—because particular circumstance is irrelevant, morality is universal. So, **the first formulation of Kant's deontological principle** is as follows: "Act only according to that maxim whereby you can, at the same time, will that it should become a universal law."[5] This rule is called the **categorical imperative**. For any action we would like to take, we must first consider what that behaviour might be like if it were a general rule of behaviour for everyone. For example, if I want to steal something from a store, I will consider this behaviour as a general moral rule; then I must consider whether I would accept this rule if it were followed universally without regard to context. I might think it's okay to steal in my circumstances, but it's logically impossible that everyone should follow that rule. Why? Because if everybody stole what they wanted, there would *logically* be no such thing as private ownership—and without private ownership, nothing could *count* as stealing. It's self-contradictory for this to be a universal law. If I could not accept a universal rule if everyone were as free as I was to act on it, then it is morally wrong for me to act on it.

Kant's belief that only an unqualified good has true moral value implies that beings who create an unqualified good are themselves a source of unqualified intrinsic value. Since humans create morality through reasoning, we are also intrinsically valuable. This leads to **the second formulation of Kant's principle**: "Act as to treat humanity, whether in your own person or in that of any other, in every case as an end and never as merely a means."[6] It is never acceptable to treat someone merely as a means to another end. You may use your waiter as a means to bring you food, for example, but you may never treat another human being as a *mere* food-bringer. This principle is sometimes called the humanity principle because it recognizes the intrinsic worth of each person and doesn't allow their value to be diminished merely for other people's purposes.

If we apply Kant's principles to our self-driving car example, it would be wrong to program a car to swerve to kill one person to save five others. The utilitarian answer has us sacrifice someone for the sake of greater happiness, or, in other words, it treats the pedestrian as a mere means to an end.

In essence these two moral principles can help answer moral dilemmas because they focus on abstract rules of behaviour. The utilitarian tells us that in the self-driving car example, we should maximize the greatest good for the greatest number, while the deontologist tells us

---

5   Kant 1785 [1993], 30.
6   Kant 1785 [1993], 36.

to respect people's intrinsic worth and to not sacrifice anyone.

Moral principles are useful for helping us make moral decisions, but as we've discovered, they can sometimes prescribe different actions. And there are several other philosophical theories of ethics that also have their own principles. Thus, ethical debate often focusses on which principles we should be acting on. When only a single principle is relevant to a scenario, or if all the principles recommend the same course of action, moral principles can be very useful for helping us make moral decisions. But when they conflict, they can also become a distraction. As we will see, there is far more to making ethical decisions than merely deciding which moral principle is the right one. But for now, we can add another tool to our toolbox: the use of moral principles to help solve moral problems.

INQUIRER'S TOOLBOX

- Ethical theories/ principles.
- Analysis of ends-means relationships.

Handout on additional moral theories.

2.2.2 Case-Based Reasoning and Moral Analogy

Another tool for making ethical decisions is to use case-based reasoning, or **casuistry**, to derive a moral judgement based upon the way similar cases have been handled or would be handled. For example, consider someone who has come upon hard times and is having difficulty feeding their family. With few alternatives, they steal a can of soup. Would such an action be morally acceptable? Normally, stealing is considered wrong, but perhaps in this case we think the need was great, soup doesn't cost much, and given their desperate circumstances, the family could use the soup more than the grocer who is trying to sell it. Given the circumstances, however, adopting such an attitude might establish a precedent that will govern how future cases are handled.

To serve as a precedent, would we have to follow a pattern of analogical reasoning where we note that, given a certain situation has certain morally salient qualities, we should judge that a certain action is ethically correct in that situation? If another situation has the same morally salient qualities, the judgement from the first case serves as a precedent for cases like it, and so we should make the same moral judgement in the second case as in the first. Imagine that, given the conditions facing this person, it was acceptable for them to steal the food. Now let's imagine another case where a person is also facing hard times and is having difficulty supporting their family. One family member becomes sick and the drug to treat them costs thousands of dollars. With no alternatives, the person breaks into the pharmacy and steals the drug. Was this morally acceptable?

The two cases may seem similar enough that we would say that if stealing is acceptable in the first case, it would be acceptable in the second. But how similar are these cases? You might be inclined to accept the theft in the first case based on the contextual need and desperation. If so, you might accept that in the second case the theft of the drugs was acceptable too. However, perhaps in the first case you only accept that the theft was contextually okay, both because of the need but also because of the small cost to the grocer. In the second case we may be more skeptical that the theft was acceptable because the item stolen is of far greater value and because the person had to break into the pharmacy to get it.

If a judge were to oversee the criminal trial of such a case, they may find more similarities to previous crimes that would inform their judgement. Perhaps this case is similar to other break-ins and the judge is more concerned about the precedent set by allowing such break-ins to occur and concludes that it was unacceptable. In other words, depending on the specific moral factors that we take to be the most significant, we will be led to compare the current case to various previous cases as a model to follow as a precedent according to which qualities we take to be the most morally relevant in each case.

INQUIRER'S TOOLBOX

- Ethical theories/ principles.
- Analysis of ends-means relationships.
- Case-based reasoning and ethical analogy.

If our judgements about what are the morally important similarities and differences between cases will determine which case we will use as a precedent, how do we determine what is a morally relevant similarity or difference? Presumably we don't want to focus on similarities or differences that are morally arbitrary to help us make our judgements.

According to philosopher Patricia Marino, "The main difficulty with case consistency is, of course, knowing what are and are not morally significant similarities and differences … judgments of moral significance are themselves also moral judgments."[7] The most significant way

---

7   Marino 2013, 742.

we can avoid being arbitrary in such judgements is to be as consistent as possible. This can involve working back and forth comparing cases, their qualities, and various principles of action to derive consistent ways to judge cases that do not resort to morally arbitrary reasons to justify different moral responses to otherwise similar cases.[8]

Thus, the use of case-based reasoning and **moral analogies** can be helpful for understanding and resolving moral questions, but they require the ability to carefully judge and describe what the moral situation is like, and doing so in a non-arbitrary way requires the ability to systematically think about cases and make comparisons. In other words, just like the use of moral principles, the application of moral precedents and analogies requires careful thinking and not simple application alone. However, we can add the use of moral analogies and case-based reasoning to our inquirer's toolbox.

3. Solving Ethical Problems in AI: How Generalizable Is Ethics?

Moral principles are convenient because they provide definitive, systematic, and universal answers to moral questions. Relying on abstract universal moral principles would allow AI developers to efficiently anticipate and consistently respond to moral concerns that might arise from the use of their product because we can start to generalize over individual cases and pick up on common moral concerns.

We might imagine that reviewing ethical issues relating to artificial intelligence could be a streamlined process. In medical contexts, researchers developing new drugs or technology must submit their proposal to ethics review boards who can vet the proposal using ethical rules and principles and determine if it is ethically acceptable to proceed. Can we simply use ethics review boards to apply moral principles to AI proposals and determine if they are ethically acceptable, or does applying ethical principles require something more?

Approaches to practical ethics that focus on the application of moral theories are called "high theory." Bioethics, the study of ethical issues emerging from biology and medicine, is strongly influenced by high

---

8   Marino 2013, 739.

theory. However, people such as philosopher Stephen Toulmin argue against relying too heavily on moral principles. His 1981 paper, "The Tyranny of Principles," recounts his experience serving as a member of an ethics review board for biomedical research, where quite often board members were able to agree (sometimes unanimously) on recommendations but would disagree about what moral principle should be appealed to. Too often, Toulmin reports, ethical problems where solutions seemed obvious to most people quickly became less temperate, less discriminating, and less resolvable once the debate turned to "matters of principle." He explains that this gives rise to an "absoluteness of moral principles that is not balanced by a feeling for the complex problems of discrimination that arise when such principles are applied to particular real-life cases."[9]

Even if using a single theory, the answer may not be so cut and dried. For example, should we try to maximize utility by focussing on each act in its own situation (what is called **act utilitarianism**) or should we adopt rules that maximize utility overall (what is called **rule utilitarianism**)? The point is that so long as we believe that context should be a factor in deciding what is ethical, then ethics is going to require more than just thinking about principles.

The problem with focussing on high theory is that if we think that there may be morally relevant specific features in individual cases, a universal moral principle will gloss over those details, potentially making you neglect important moral factors. You may become so focussed on the principle that you believe is at stake, that you become less critically minded about how that principle should be applied and less attentive to other problems. As Toulmin notes, "Oversimplification is a temptation to which moral philosophers are not immune, despite all their admirable intellectual care and seriousness; and the abstract generalizations of theoretical ethics are ... no substitute for a sound tradition in practical ethics."[10]

3.1 THE LIMITS OF MORAL PRINCIPLES AND AI

In his 2019 paper "Principles alone cannot guarantee ethical AI" Brent Mittelstadt considers the applicability of high theory to AI. He notes that there are already 84 initiatives by different groups articulating

---

9   Toulmin 1981, 32.
10   Toulmin 1981, 31.

ethical principles for AI who seek to translate these principles into more specific governance frameworks and professional ethics codes.[11] Most of these initiatives, such as the European Union's High-Level Expert Group on AI, have converged on principles such as fairness, prevention of harm, explicability, and respect for human autonomy.[12] These principles are supposed to embed normative consideration in technology design and governance and function like high-level principles in bioethics. Nevertheless, Mittelstadt notes important differences between bioethics and AI that make the use of principles more complicated.

Unlike in medicine, where there is a common recognized aim (to promote the health and well-being of the patient), AI development is largely driven by private sector aims such as cutting costs or increasing profit. The fundamental aims of developers, users, and affected parties are not the same. Because there is no patient whose interests would have the highest priority, there is also no relationship of trust. A doctor acting on a patient's behalf for the patient's best interest is involved in a relationship of trust—that is, has certain **fiduciary responsibilities** to that patient. AI developers, by contrast, have no recognized fiduciary responsibilities to the public or to public interests over private ones. This makes it more difficult to develop commonly understood and agreed upon sets of principles to govern the field.

In addition, AI development lacks a long history of professional development or a singular professional culture. There is no sense of what makes a "good" AI developer when compared to a profession like medicine, which has a wealth of standards for what would constitute a "good" doctor. Such professional cultures, "provide a historically sensitive account of the obligations of the profession against which negligent content and practices can be identified."[13] They make it easier for high theory to work because the principles provide a common language to deal with practices where ethical necessity is recognized.

AI development has no common professional culture. Creating ethical frameworks that cover all ethical concerns is challenging when "the impact of decisions taken in designing, training, and configuring AI systems for different uses may never become apparent to developers."[14]

---

11   Mittelstadt 2019, 501.
12   European Commission 2019.
13   Mittelstadt 2019, 502.
14   Mittelstadt 2019, 503.

Often in AI development no single person has a full understanding of the system's functions and being unable to understand the impact of ethical decisions makes it difficult to create ethical norms for the profession.

Broad principles like "fairness" or "respect for human dignity" are often too abstract to provide meaningful guidance. What one developer might consider "fair" another might consider grossly "unfair." As Mittelstadt notes, "statements reliant on vague normative concepts hide points of political and ethical conflict. 'Fairness', 'dignity' and other such abstract concepts ... have as many possible conflicting meanings requiring contextual interpretation through one's background political and philosophical beliefs."[15]

As Mittelstadt explains, "Professional societies and boards, ethics review committees, accreditation and licensing schemes, peer self-governance, codes of conduct, and other mechanisms supported by strong institutions help determine the ethical acceptability of day-to-day practice by assessing difficult cases, identifying negligent behaviour, and sanctioning bad actors."[16] These institutions help translate principles into practice by studying, testing, and revising recommendations and norms over time by considering different cases. The field of AI development lacks similar institutions to translate principles into practice. Translation involves "the specification of high-level principles into mid-level norms and low-level requirements," although "norms and requirements cannot be deduced directly from mid-level principles without accounting for specific elements of the technology, application, context of use, or relevant local norms."[17] As we will discuss in future chapters, for example, it can be difficult to articulate the meaning of privacy as a high-level concept and then translate that into practical guidelines for AI development, and even more difficult to evaluate those guidelines given the potential risks and trade-offs when they are put into practical use. So, to return to our earlier question, clearly all the ethical problems of AI cannot be solved by simply relying on ethics review boards that apply broad ethical principles.

---

15   Mittelstadt 2019, 503.
16   Mittelstadt 2019, 503.
17   Mittelstadt 2019, 503.

3.2 DOES ETHICS REQUIRE INQUIRY?

Ethics involves much more than the development and application of moral principles or moral precedents. These are useful but limited. Ethics requires considerations of context, use, logistics, and resources. A similar issue occurs even in the sciences. Philosopher of science Nancy Cartwright has argued that the laws of physics are literally false if taken to be universal because they express idealized relationships that we don't find in the real world.[18] In order for a law of physics to be applicable, it requires mediating models to bridge the gap between the theoretical and the observed world. Thus, the usefulness of a scientific law will depend on how well the scientist is first able to model the situation where they would like to apply the law.

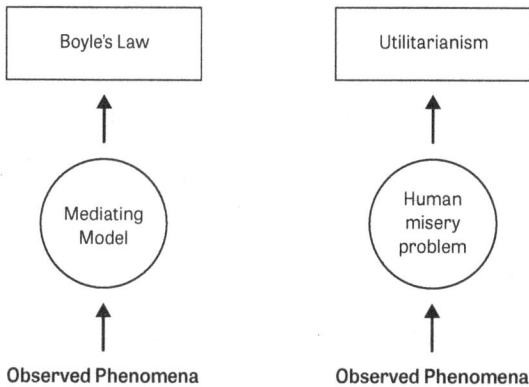

FIGURE 2 · Application of a scientific law requires that observed phenomena are conceptualized in such a way as to make it applicable. Mediating models allow for this, and in ethics we too need to conceptualize and describe problems in just such a way as to make theories and principles applicable.

Perhaps the same is true in ethics as well. To understand how to use ethical principles we must also appraise the situation we are facing to determine how applicable the law is (Figure 2). The application of ethical principles requires careful attention to the particular characteristics of situations. Thus, it may be better to focus on a broad approach to studying ethical questions by focussing on moral inquiry as a process.

---

18   Cartwright 1983, 2.

The tools we have so far in our inquirer's toolbox are useful for ethical thinking but remain limited. To appreciate the different perspectives, interests, goals, and logistics involved in AI development, we will need to make inquiries into specific situations that call for our ethical attention. We need to supplement the tools in our toolbox by considering a general framework for moral inquiry and discovering some useful ways to analyze ethical problems and experimentally assess potential solutions with reference to real-world conditions.

4. Building a Framework for Moral Inquiry

Let's recap the discussion about ethics. Ethical thinking and ethical concern follow from the things that we care about. But ethics involves making choices between incompatible options. We don't want self-driving cars to hit anyone, but we may have to accept that they do so in order to save others. An AI developer may not want to create an algorithm that discriminates against poorer people, but credit scores are a desirable and efficient piece of data.

If ethics begins when we are faced with a choice between incompatible ends that we value, then it highlights the importance of paying attention to the way that the world is and how we can act in it, and not just how we would like to act in it. As we discovered when thinking about ethics review boards, ethical action in the world requires analyzing information, considering logistics, solving problems, and trying out different solutions in different contexts. Hence an account of ethical inquiry must make room for such considerations.

To articulate an account of moral inquiry, we will need a better understanding of inquiry as a critical reflective process in general. Thus, we will consider a general account of inquiry as a reflective and logical process using philosopher John Dewey's 1909 work *How We Think* because it focusses on the practical ends of inquiry, such as making a value judgement, and on the collection and critical reflection of facts needed to make such judgements.

4.1 THE PURPOSE OF REFLECTIVE MORAL JUDGEMENT

If ethical judgements are about deciding how to act in the world when faced with incompatible ends that are both valued, then ethical judgement will involve determining which of these ends is of greater value. As John Dewey explains "Moral judgments, whatever else they are, are a species of judgments of value."[19]

Value in its unreflective sense is, according to Dewey, a kind of behaviour or a disposition to behave in certain ways. When we value something, we esteem it, we pursue it, or approve of it. You are going to put energy into either obtaining it or keeping it once you've obtained it.[20] Hence **valuing** could be described as an unreflective form of value. However, not all values are unreflective. As discussed in Section 2.1 you can also engage in reflective thinking and make a **value judgement**, which will consider various valued ends in relation to the means required to obtain them. A value judgement involves appraising things we value; thus, it will not be concerned with what we *do* value, but with what we *should* value.

Recall how consideration of the means available to help achieve an end in relation to other possible ends presents us with the ability to critically reflect on what we value. To think "is to look at a thing in its *relations* with other things, and such judgement often radically modifies the original attitude of esteem and liking."[21] We might learn to avoid short-term pleasures, for instance, if they include long-term pain. If so, it is because of critical reflection that we understand how those short-term pleasures relate to long-term pain.

This allows us to understand, for example, how despite its unpleasantness, getting a flu shot can ultimately be a good thing. As Dewey explains, "All growth in maturity is attended with this change from a spontaneous to a reflective critical attitude. First, our affections go to something in attraction or repulsion; we like and dislike. Then experience raises the question whether the object in question is ... such [as] to justify our reaction to it."[22]

---

19   Dewey and Tufts 1932, 290.
20   Dewey 1939, 14.
21   Dewey and Tufts 1932, 291.
22   Dewey and Tufts 1932, 291.

We shouldn't confuse making a value judgement with mere valuing. It is too easy to reduce ethical claims and judgements to mere subjective preference or to say that all values are subjective. We can forget that some things we desire are only the result of critical thinking, problem solving, and experimentation. If we take all values as mere subjective desires, we undermine the purpose of moral inquiry. Through a process of critical assessment of the things we value by considering them in relation to other things, we begin to criticize and refine what we value until we reach a decision about what is the *best* way forward. Therefore, the purpose of a moral judgement is to determine the value of various incompatible pathways before us when we are unsure how to proceed. We come to a moral judgement through a process of reflection about means and ends that we can call moral inquiry.

4.2 THE FELT DIFFICULTY

Moral inquiry begins where there is tension between incompatible ends. This may take the form of desiring a certain end and lacking the necessary means required to obtain it. Inquiry may then begin with a focus either on finding suitable replacement ends or finding different means for obtaining the desired end. Alternatively, moral inquiry could also begin because we wish to adopt a moral belief that is incompatible with other known facts and we need to resolve that tension. Either way, inquiry begins with what Dewey calls "a felt difficulty" caused by this tension, and the primary purpose of inquiry is to resolve this tension and adopt a new belief (Figure 3).[23]

Another way of understanding a felt difficulty is for you to think of something that is valued but doubt that it should be. When we find ourselves in such situations, we tend to describe them as "doubtful," "ambiguous," "confused," "conflicting," or "obscure."[24]

As Dewey explains, "Thinking ... does not occur just on 'general principles.' There is something specific

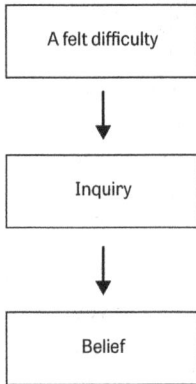

FIGURE 3 · A felt difficulty prompts inquiry and that inquiry will only be resolved when we adopt a belief that resolves the difficulty.

23   Dewey 1909, 72.
24   Dewey 1939, 105.

which occasions and evokes it."[25] Inquiry begins when we are faced with a situation that causes an interruption to our normal thoughts and actions that undermines our beliefs, creating a state of perplexity, hesitation, and doubt. The purpose of inquiry is to eliminate this state of perplexity and adopt a belief that we assent to in a way that is stable and that keeps us from having to return to inquiry.

An important insight of this relationship among a felt difficulty, inquiry, and belief is that the difficulty that prompts us to inquire has a role in regulating inquiry and its conclusion. The nature of the problem, for example, will inform what we consider relevant to a solution. In ethics, we can never directly test our conclusions against cosmic ethical standards. However, we do have the ability to test whether our conclusions resolve the difficulties that prompted moral inquiry in the first place or if they've left us more frustrated and confused. If our conclusions only cause further problems, we will be forced to return to the process of inquiry until we find an answer that settles the moral question in a way that satisfies the difficulty and allows us to move on to other things.

If we recall our earlier dilemma over whether we should help a cousin move or go on vacation with a friend, we can understand how such a felt difficulty would affect our moral inquiry. Any potential solution to that problem will involve considerations of timing, location, how much needs to be moved, how we plan to move it, whether there will be hurt feelings, whether one person will be worse off if I break my promise, and so on, for these are the factors which are in conflict. As Dewey explains, "A question to be answered, an ambiguity to be resolved, sets up an end and holds that current of ideas to a definite channel. Every suggested conclusion is tested by its reference to this regulating end, by its pertinence to the problem in hand.... *The problem fixes the end of thought and the end controls the process of thinking.*"[26]

Just because our conclusions resolve our difficulties doesn't mean that our conclusions will be personally satisfying. The moral conflicts that prompt inquiry won't go away simply by finding the answer we are happiest with. Our conflict is partially based on the way the world is, and so there is always a risk that we will be tempted to prematurely end our

---

25   Dewey 1909, 12.
26   Dewey 1909, 11, Dewey's emphasis.

inquiry when we find a conclusion that we like rather than one that will actually solve our problem.

You might convince yourself that your cousin doesn't need much help moving and it's acceptable to break your promise, but that doesn't change whether your cousin actually needs help or how they will feel when you don't help them. Similarly, if you do still have misgivings about your conclusion, it's important that you follow up on those concerns and if need be, they can be referred to a community of fellow moral inquirers for review. If we falsely convince ourselves of the efficacy of our own solution, we will be more likely to make mistakes or to have to return to inquiry to rethink the problem.

4.3 PATTERN OF MORAL INQUIRY

4.3.1 Defining the Problem

A medical patient presents with various phenomena from which the doctor will try to determine the patient's symptoms. If the patient presents with a simple bleeding issue, diagnosing the patient may seem straightforward. Alternatively, if the patient also seems delusional, the doctor must assess if there might be a connection with the bleeding issue. Is the delusion a psychological issue or a neurological issue? The symptoms that the doctor believes that patient has will define the scope of what the doctor needs to consider for a diagnosis. Different symptoms will suggest different illnesses and different things to look for. If the doctor misunderstands the symptoms, they misunderstand the problem and must reconsider how they have defined the patient's illness.

This is an important lesson for all inquiry. If we don't understand the nature of the problem, we won't know how to fix it. We must begin by trying to get a grip on the nature of the problem that we seek to resolve. Often (but not always) it is unclear exactly what the moral problem is, and such problems don't come with labels. As Dewey explains, "Inquiry may be regarded as a request 'for information.' But the needed information is not handed out ready-made by nature. It requires judgment to decide what question should be asked ... since it is an affair of formulating the best methods of observation, experimentation, and conceptual interpretation."[27]

---

27   Dewey 1938, 170.

There may be different ways to define ethical problems that may include or omit information for consideration. People with different perspectives may define problems in different ways. During the COVID-19 pandemic, for example, governments all over the world had to try to define the nature of the problem in terms of solutions. Should we focus on contact tracing and quarantine, or should we focus on vaccination? If we do focus on vaccination, is the problem about getting the entire population vaccinated or is it just about keeping the healthcare system from being overburdened? Different problems suggest different goals, and thus different information relevant to those goals.

When we inquire, we must evaluate information to determine whether or not it is relevant to the problem and a potential solution. It will never merely be a question whether information is factual, but whether it is relevant. If a moral problem is defined too broadly it will be more difficult to detect the specific causes of the problem and may lead to overly general solutions. Alternatively, defining the problem too narrowly means that we are more likely to miss out on important clues and information or to exclude things that ought to be considered. If the problem is badly defined, inquiry will struggle and we will likely be forced to start again, but if the problem is well defined, the solution may be far more apparent. As Dewey notes, "It is a familiar and significant saying that a problem well put is half-solved. To find out what the problem [is] ... is to be well along in inquiry."[28]

Let's consider an example. It is revealed that the company that you work for uses an algorithm to aid in hiring practices, and this process is discovered to have several biases. Merely defining the problem as a "biased algorithm" is too broad since it is unclear what kinds of biases there are or what kinds of moral problems they create. Some forms of bias can also be good or reflect the intentions of the designer. Alternatively, not all biases affect people in the same way. If we attempt to simply reduce bias overall, we may find that our efforts are less efficient or less effective than if we had attempted to determine the specific kinds of biases that cause the most disproportionate harm to specific groups. A badly defined problem will lead to poorly thought-out solutions.

---

28   Dewey 1938, 108.

4.3.2 Suggested Solutions?

With a tentative definition of the problem, ideas can present them-selves as possible solutions. Depending on the nature and complexity of the problem, we may pull from several sources to inform our think-ing. For example, if we are faced with an ethical dilemma where there seems to be a conflict of values relating to autonomy, we may choose to consult deontological ethical principles and theorists for their insight. Alternatively, the conflict might be avoided by paying attention to logis-tics. In that case, we might pull information from the experiments of engineers and scientists.

Suggested solutions will be helpful only if the problem is well defined. As discussed earlier, moral analogies can be useful at this stage because if we can find similar problems where the answer is clearer, it may help us find an answer in this case as well. However, as noted, judgements about morally relevant similarities and differences are themselves moral judgements. At this stage in inquiry our defined problem and any suggested solutions must be tentative. Suggested solutions are merely hypothetical until they can be critically consid-ered and tested. Our problem might be badly defined, and we must go backwards and rethink it.

We can also use moral principles to help propose possible solutions based on the way we understand the problem. However, as we noted, moral principles can often conflict with each other, and the prescrip-tions they might offer don't always account for the logistical difficulties and conflicts involved with trying to carry them out. Moral principles present specific perspectives on morality that might prove useful, but they can also be limiting. It might make us focus on certain things that the principle identifies as useful, but we might also ignore parts of the problem that the principle considers irrelevant that might be important for a solution.

Moral principles can tell us which aspects of a moral problem are morally important, but rarely do they capture every aspect of moral thinking that we tend to think are relevant. As Dewey explains, "Morals is not a catalogue of acts nor a set of rules to be applied like drugstore prescriptions or cook-book recipes. The need in morals is for specific methods of inquiry and of contrivance: Methods of inquiry to locate dif-ficulties and evils; methods of contrivance to form plans to be used as working hypotheses in dealing with them."[29] This is why it isn't a good idea to simply choose whatever moral theory most closely aligns with your own tastes because your own biases can affect your thinking.

Dewey's advice for using moral principles in reflective moral inquiry is to "look upon all the codes as possible *data*; it will consider the conditions under which they arose; the methods which consciously or unconsciously determined their formation and acceptance; it will inquire into the applicability into present conditions. It will neither insist dogmatically upon some of them, nor idly throw them all away as of no significance."[30]

4.3.3 Reasoning

A suggested idea can seem good until it is considered in more detail. Often this will involve real-world application of the idea, or imagining what the idea would entail if acted upon. Both processes are necessary for refining our suggested solutions. Dewey explains, "Not every suggestion is an idea. The suggestion becomes an idea when it is examined with reference to its ... capacity as a means for resolving the given situation."[31] Reasoning about a potential solution might begin with seeing if the solution really fits with all the facts of the case, or determining its appropriateness given what we have affirmed so far in inquiry.

Reasoning about suggested solutions involves deliberation, determining what it would be like if the suggested solution were acted upon. Dewey explains, "Deliberation is an experiment of finding out what the various lines of possible action are really like."[32] In the example of helping our cousin with moving: if we hired a mover, we need consider what that would actually be like. Will they get there in time to help your cousin? Will you have the money to pay them? Will your cousin be happy with your solution? We imagine such a scenario to determine how suitable any suggested solution might be.

Because deliberation is a process of experimentally trying to reason out the consequences of our solutions, the reasoning stage of inquiry will naturally be drawn to analysis of ends-means relationships. Recall that an end is a goal or settlement of our problem while the means are

---

29   Dewey 1922, 170.
30   Dewey and Tufts 1932, 191.
31   Dewey 1938, 110.
32   Dewey 1922, 190.

those logistical factors required to obtain that end. According to Dewey, we can deliberate about which ends are more valuable by considering that end, both in terms of the means required to obtain it and the consequences that the end will likely produce given the means required to obtain it, or what he calls an **end-in-view** (Figure 4). In evaluating potential ends, "we can judge its nature, assign its meaning, only by following it into the situations wither it leads, noting the objects against which it runs and seeing how they rebuff or unexpectedly encourage it."[33]

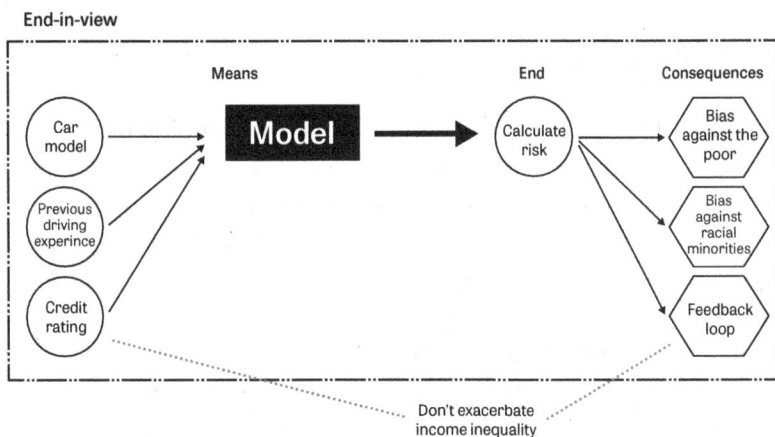

FIGURE 4 · Using Dewey's concept of an end-in-view, we can consider an end in terms of the means used to achieve that goal and the consequences it produces. Using credit ratings as a means to decide insurance rates leads to consequences that conflict with other ends-in-view that we have. This not only helps us understand the nature of the moral conflict that prompts inquiry, but it also helps us figure out what we can do to resolve it.

This way of thinking represents a kind of moral maturity. If we are faced with a conflict between things that are valued, it would be immature to choose based simply on which thing we impulsively want more. A child might impulsively say that they only wish to eat candy simply because it tastes good. The difference between the child and the adult who rejects the candy-only diet is that the adult understands the consequences of only eating candy and thus understands the meaning of such a proposed action far more clearly. The adult understands a fuller ethical

---

33   Dewey 1922, 192.

meaning of a candy-only diet as an end-in-view, whereas to the child it is merely a preferred end.

When we consider suggested solutions as ends-in-view we not only gain a better understanding of what it would mean to act on a proposed solution, but we gain a means of rationally comparing and criticizing it. Dewey explains that when we consider an end, if we find that it would take too much time, too much energy, or if it were attained this would bring about too many additional problems or conflicts with other things we value, we will judge that it is a bad end.[34] The fact that eating only candy means you will likely get sick makes it a bad idea. Things that can seem or feel good or valuable can turn out to be bad in practice. Similarly, things that can feel bad or undesired can ultimately be good. We may not enjoy getting a vaccination shot, for example, but we know the pain is ultimately *worth* it.

By tracing our valued ends as ends-in-view, we get a more specific understanding of that valued end, which makes it easier to appraise, particularly when comparing ends-in-view to each other. When we need to evaluate competing ends-in-view, we not only begin to anticipate which kinds of ethical solutions will have the best practical chances of resolving our ethical problem, but we also understand the competing trade-offs that each potential solution carries, and how they may create obstacles for us. Understanding how best to achieve an end, or the conflicts we will run into in pursuit of this end, will also help us come up with means of experimentally testing whether a proposed solution will work out as anticipated, and if suggested solutions are actual solutions. By reasoning out our potential solutions as ends-in-view, we might also find ways of modifying or combining them for better effect.

4.3.4 Experimentation

Once we reason out potential solutions, we must experimentally test and verify them. Testing may lead to new observations and new information, potentially causing us to rethink the entire problem. Importantly this means that inquiry is not one-directional. We may find ourselves going back and forth from observing to defining the problem to reasoning out solutions and experimenting only to find that our assumptions

---

34   Dewey 1939, 24.

were faulty, and we need to rethink the problem considering new observations (Figure 5).

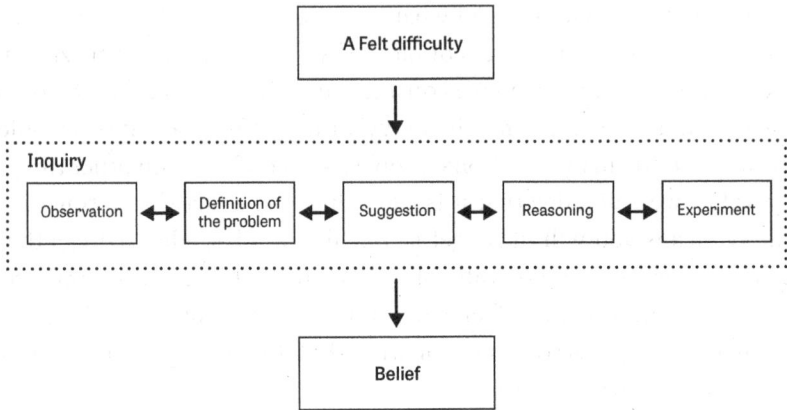

FIGURE 5 · Inquiry begins with a felt difficulty and ends with the adoption of a belief with the process of inquiry consisting of several stages. Sometimes we may have to go back and forth from one stage to the next in pursuit of an answer. (*Diagram adapted from Brown 2012.*)

Once we have a hypothetical solution, we must test it in practice. Using experimental trials, noting both where we fail to bring about the ends desired and where we succeed, where certain conditions make certain ends possible, we discover through intellectual means the creation of desires which are more suitable given the conditions available for realizing them. If your proposed solution doesn't work, your problem will remain (or may even be potentially worse). If your solution does work, then it will resolve your problem as you understand it and eliminate your felt difficulty.

4.4 VALUE JUDGEMENT AND BELIEF

There may be multiple ways to resolve real world ethical problems and some problems may call for novel experimental approaches. We can imagine, for example, why many ethical principles and precedents will have difficulty being applied to issues involving artificial intelligence because the field presents so many novel issues that previous ethical frameworks didn't need to consider. Those engaging in reflective moral inquiry should be "on the outlook for something better. The conscientious person not only uses a standard in judging but is concerned to revise and improve his standard."[35]

As mentioned, if your hypothetical solution works out as anticipated, you will arrive at a firm belief about how that problem should be resolved. With the moral conflict resolved, the need for inquiry ends. However, sometimes solutions that seem to work at first might end up causing unanticipated problems. Confirmation that your solution was best suited to resolve your problem may not come all at once or only by you.

Remember that we all have our own biases, and so even if we think our difficulty is resolved, it might only be because someone else hasn't yet pointed out the flaws in our reasoning or the problems that we created that we didn't even notice. That's why, as Dewey explains, "An inquirer in a given special field appeals to the experiences of the community of his fellow workers for confirmation and correction of his results. Until agreement upon consequences is reached ... the conclusions that are announced by the individual inquirer have the status of an hypothesis."[36]

In other words, while the solutions to some moral problems might be obvious, other moral problems may take years to work out or may never be fully worked out. Progress in finding solutions, however, will not be found by changing how a single inquirer feels but by acting in the world and understanding the feelings of others. This is why moral thinking requires critical thinking in the form of moral inquiry.

5. Conclusion

Ethics can be complicated. We might choose to take one of two attitudes when faced with an ethical dilemma. We could react non-reflectively and simply grasp at any solution that seems good enough or we could react reflectively and critically examine the problem. The history of ethical thought is filled with principles, theories, and precedents that can help us find answers, but we must always remember to ask ourselves where they came from, whether they are still applicable given novel conditions, whether they might need modification, or whether we might need an entirely new approach.

---

35   Dewey and Tufts 1932, 301.
36   Dewey 1938, 490.

It will always be tempting to prescribe simple solutions to complex questions and fall back on preferred precedents or principles. However, rigidly following a moral theory because you like it best, and not because it is suited to the situation, may only make the problem worse. We must be prepared to subject all our moral understandings, views, and judgements to reflective criticism.

Now that we have considered the steps of moral inquiry along with forms of moral thinking that can be useful for inquiry, we can try to convert what we've learned into a helpful toolbox for anyone engaging in any form of moral inquiry. In the remaining five chapters, we will be applying our process of inquiry to several problems and areas of moral tension that have arisen concerning AI. This will include unpacking the means-and-ends relationships that are at the root of such problems, but also considering suggested solutions in each case as well. As our investigations into ethical issues involving AI continue, we will have opportunities to add additional questions to help stimulate inquiry.

INQUIRER'S TOOLBOX

1. How have I defined the ethical problem?
2. Am I defining this problem too broadly or too narrowly?
3. What conditions give rise to the problem I am facing? Can they be mitigated?
4. Am I considering all the information relevant to a solution? Is that information reliable?
5. Which moral theories or principles might be helpful to consult in this case? Are there areas where the theory may be irrelevant or unhelpful?
6. When I consider a moral solution as an end-in-view, does it make sense? Is it a practical workable solution?
7. How would I test any assumptions I have regarding the nature of the problem or a hypothetical solution?
8. When I consider how chosen ends might function as means to future ethical situations, are there major ethical concerns to consider?
9. Are there historical precedents or analogies that can inform how I might understand a problem or a potential solution? In what ways is that precedent helpful or relevant? In what ways is it not?
10. Is my judgement coherent? Is it justified given other judgements I have made about the problem or similar problems?
11. Are there biases or limitations on my perspective which might require insight from others? Am I cherry-picking information, cases, principles, theories, solutions?
12. Are there multiple solutions to this ethical problem?

ADDITIONAL MATERIAL

Artificial Intelligence • the ability of machines to engage in problem solving.

Algorithm • a set of instructions used to define an operation that takes in an input and produces and output. The exact meaning of the term can vary across the AI and machine learning field.

End • a possible goal or settlement point for a dilemma, one of two elements of a value judgement.

Means • one of two elements of a value judgement, the practical and logistical factors that are required to obtain an end.

Utilitarianism • a moral theory which holds that all actions should aim at maximizing utility, that is, promoting happiness and reducing pain.

Intrinsically valuable • something that is considered valuable in and of itself without regard to its instrumental uses.

Principle of utility • holds that you should maximize the greatest amount of happiness for the greatest number of people.

Deontology • an ethical theory which holds that an action is right if it accords with our ethical duties.

Categorical Imperative • a moral duty we must follow, regardless of all our desires and of context.

First formulation of Kant's principle • "Act only according to that maxim whereby you can, at the same time, will that it should become a universal law."

Second formulation of Kant's principle • "Act as to treat humanity, whether in your own person or in that of any other, in every case as an end and never as merely a means."

Casuistry • a type of reasoning that derives moral judgements from a case study and seeks to apply those judgements in other similar cases; also called case-based reasoning.

Act utilitarianism • a branch of utilitarianism that seeks to maximize utility with each individual action.

Rule utilitarianism • a branch of utilitarianism that seeks to maximize utility through a set of moral rules that produce more good than bad overall.

Fiduciary responsibilities • a duty of care owed to a beneficiary to act in their best interest.

Valuing • an unreflective form of value based on our impulses and desires.

Value judgement • a reflective form of value that involves appraisal of what we should value, including the means required to obtain the things we value.

Moral analogies • the use of analogical reasoning to help inform our ethical thinking and to draw comparisons.

End-in-view • the consideration of an end/goal in terms of the means required to obtain it and the consequences it will likely produce.

---

1. Why can't we just make algorithms and AI-powered technologies to be ethical? Reflecting on the material from this chapter, why can't we program algorithms to be ethical? Why isn't it this simple?

2. Utilitarians such Bentham argue that ethics is about the consequences and maximizing pleasure (or minimizing pain), while deontologists contend that ethics is about duties, intentions, and rationality. What are the limits of each perspective?

3. Many ethical theories discuss ethical acts in terms of what would be an ideal ethical response to an ethical problem. Why can it be limiting to think about ethics only in terms of what is ideal?

4. The first formulation of Kant's deontological principle reflects what duties we have to one another as rational beings. However, some find this principle too restrictive and open to counterexamples. Can you think of any ways of questioning this criticism, and do you agree that it is too rigid? Explain.

5. The second formulation of Kant's deontological principle states that we should never treat one another *merely* as a means to our ends. What role does the word "merely" or "only" play in this principle?

2

The Moral Responsibilities of Scientists

Many scientists and researchers tend to value truth and knowledge for its own sake, and this value is sometimes given supreme importance in their pursuits and studies. However, what should our response be to research that threatens our democracies or our families, or research that may (inadvertently) lead to human rights abuses, or that makes us less free and more unhappy?

In 2018, four researchers published a paper titled "Expression of Concern: Facial feature discovery for ethnicity recognition" in the journal *Data Mining and Knowledge Discovery*. The "concern" here was not ethical, but rather about the efficacy of previous systems for ethnicity recognition: it suggests other methods. The following year, several researchers wrote to the journal's publisher, Wiley, to retract the article. Wiley denied the request, on the grounds that the objections raised were not to the methodology of the research described in the article, but rather to its practical use. The paper describes the researchers' success in getting algorithms to identify a person based on their ethnicity according to facial characteristics.[1]

The paper posed a moral concern because it included datasets of Chinese Uyghur, Tibetan, and Korean students who attended Dalian University in China. China has often been condemned for its heavy

---

1   Wang et al. 2019.

surveillance and mass detention of Uyghur Muslims, and so researchers who believed that it was disturbing that academics would try to build such algorithms called for the article to be retracted. In response, Wiley defended the article, arguing "We are aware of the persecution of the Uyghur communities ... However, this article is about a specific technology and not an application of that technology." Meanwhile, many scientists have gone a step further and have argued that all unethical biometric research should be retracted.[2] They urge the scientific community to take a firmer stance against unethical facial-recognition research by not only denouncing controversial uses of technology, but the research foundations as well, calling on researchers to avoid working with firms or universities linked to unethical projects.[3]

This case presents several important ethical questions:

1. Can we really make a clean distinction, as Wiley attempts, between the research of a technology and the application of that technology?
2. Should scientists and AI developers be more conscious of the ethical implications of their work?
3. Do they have the responsibility to refuse to engage in such research at all?
4. Is there an ethical line between research and the application of research when it comes to AI?
5. If the Chinese Government did use the researchers' methods to persecute the Uyghurs, would those researchers be morally responsible for the harms caused by their research?

To attempt to answer these ethical questions, we can consult our inquirer's toolbox, in particular questions 3 and 9 in the previous list. To inquire into the ethical responsibilities of scientists and gain a fuller understanding of how the relationship between pure and applied science has been understood, we must consider the historical context that drives discussion about the moral responsibilities of scientists. Thus, we will need to consider some history and philosophy of science to appreciate the origins of how scientists came to understand their

---

2   Moreau 2019.
3   Noorden 2020.

moral responsibilities, how this view has changed over time, and what sorts of moral responsibilities scientists and engineers should have if their work is going to radically change society. Through this inquiry, we will be better able to determine if AI developers should be held ethically responsible for their efforts and if new discoveries in science, such as AI, will always be worth their ethical costs.

INQUIRER'S TOOLBOX

3. What conditions give rise to the problem I am facing? Can they be mitigated?

9. Are there historical precedents or analogies that can inform how I might understand a problem or a potential solution? In what ways is that precedent helpful or relevant? In what ways is it not?

1. Pure vs. Applied Science and the Linear Model of Scientific Development

Wiley's defence for not retracting the article was that it was about the research and development of techniques for facial recognition and not about what such research could be used for. This is in keeping with a common distinction made between **pure science** and **applied science**. Whereas applied science is about the application of scientific theory for the purposes of achieving some specific goal, pure science is supposed to be about pursuing the truth about nature for its own sake. Albert Einstein discovered that matter can be converted to energy, but this kind of research is different from the kind that describes exactly the material necessary to create a chain reaction for fission. One kind of research seeks to understand how atoms work, while the other kind seeks to understand how to make atoms work for us.

This division between pure and applied science dates to the time of Francis Bacon. As a champion of a new method during the scientific revolution, Bacon distinguished between what he called experiments of light (*experimenta lucifera*) and experiments of fruit (*experimenta fructifera*). "Light" and "fruit" here are metaphors, the former meaning *truth* and the latter *results*. While the latter are experiments that serve some immediate purpose, the former are experiments that serve no immediate use. As he explains, "our hope of further progress in the sciences will then only be well founded, when numerous experiments shall be received and collected into natural history, which, though serving no use in themselves assist materially in the discovery of causes and axioms; which experiments we have termed enlightening, to distinguish them from those which are profitable."[4]

In his *Novum Organum* (1620), Bacon explains there are various temptations which can distort our thinking and make us less scientific, like the desire to profit from our research before it is ready. So, it is better in general to pursue science "for its own sake" rather than for our short-term ends. Robert Proctor describes Bacon's view: "the power of science is not to be gained directly, or immediately, but only through the patient accumulation of knowledge through experiments."[5] This would guide the distinction between pure and applied science, but as Proctor argues, this arrangement served a social purpose as well:

> In the 'royalist compromise' of seventeenth-century English science, natural philosophers of the Royal Society of London promised not to meddle in matters of 'Divinity, Metaphysics, Moralls [or] Politicks,' in exchange for rights to publish without censorship, to correspond freely with other members, to pursue science with the blessing and support of the state. Neutrality was a bargain struck, the price science had to pay, for social legitimation in the eyes of church and state.[6]

Pure science invested in the pursuit of knowledge for its own sake was decoupled from social, ethical, and religious concerns, which meant that science would not be a threat to matters of state.

Bacon's position that science should be a morally neutral pursuit of truth for its own sake, which eventually will lead to societal and technological progress, is today called the **linear model of scientific development**. It holds that investments in pure science (or basic research) with the freedom to pursue knowledge for its own sake will eventually yield findings that aid in applied research and development necessary for technological and economic innovation. For example, in 1895, German physics professor Wilhelm Röntgen was experimenting with the use of cathode rays, which are streams of electrons produced in a vacuum tube. After wrapping the tube in black paper to prevent visible

---

4   Bacon 1902, 80.
5   Proctor 1991, 33–35.
6   Proctor 1991, 7.

light from escaping, he discovered that a screen painted with barium platinocyanide became fluorescent due to a yet unknown kind of ray.[7] Knowing that the ray could pass through certain materials, Röntgen experimented until he found a medical application after making an image of the bones in his wife's hand. He discovered the x-ray.

This discovery would revolutionize medicine with the beginnings of radiology, not to mention additional applications of x-rays in fields like astronomy and physics. Notice the relationship between society and science. "Give us the freedom to pursue truth in a morally unburdened way and it will eventually pay off for the rest of society in the form of applied science," the advocate of such a position would claim.

Many scientists and philosophers have advocated this position. In a 1948 article for the *Bulletin of the Atomic Scientists* (the scientific organization you may know today for their Doomsday Clock), physicist Percy Bridgman argued for the need for freedom from moral responsibilities for the pursuit of truth. Bridgman wrote: "The challenge to the understanding of nature is a challenge to the utmost capacity in us. In accepting the challenge, man can dare to accept no handicaps. That is the reason that scientific freedom is essential and that artificial limitations of tools or subject matter are unthinkable."[8] Philosopher Hermann Lübbe believes that "Free research, itself legitimized traditionally by recourse to curiosity, remains, in the long run, the only relevant and usable kind of research," but to do this science must enjoy "a morally unencumbered freedom from permanent pressure to moral self-reflection."[9]

Similarly, even if a scientist is working on applied questions in one area, they may unintentionally discover applications of experiments that can profoundly alter society. Consider the example of William Henry Perkin, who in 1856, while attempting to synthetically produce quinine, accidently discovered the first synthetic dye. The result would revolutionize the textile industry. In fact, the development of chemistry and its application to textiles helped create the modern chemical industry as we know it.[10] In these cases we note that science can offer beneficial yet unintended consequences in application; there is thus a social benefit to allowing scientists great latitude to experiment.

---

7   Rosenbusch and De Knecht-Van Eekelen 2019, 89.
8   Bridgman 1948, 72.
9   Lübbe 1986, 82.
10   Aftalion 2001.

If we consider the Wiley case again, we can better understand Wiley's position. The article demonstrates a form of pure science to the extent that it seeks to uncover truths about the nature of artificial intelligence and above all is only a demonstration of a method. The paper itself does not speak to issues of government or social policy concerning its usage. Thus, it would be wrong to infringe on scientific freedom by retracting the article because the researchers are simply seeking to expand our knowledge.

But this isn't the end of our inquiry, for as we discussed in previous chapters, inquiry often leads to new questions. Is the pursuit of truth always more important than other things we care about? What if the benefits from basic research aren't worth the cost? Just how free from moral responsibilities should scientists (and AI developers) be?

The success of the scientific revolution, followed by its many benefits to those societies which invested in science, seemed to demonstrate the wisdom of the linear model and the pursuit of pure research, and the benefits of scientific freedom. However, this debate radically changed when industrial mass production and the chemical warfare of World War I unleashed tragedies and horrors never seen on such a scale. Many began to question the wisdom of whether the linear model would naturally produce "innovations" that would benefit society, as so many of its adherents seemed to promise.

2. Fritz Haber, Bringer of "Bread from the Air," Father of Chemical Warfare

Prior to 1900 the planet's capacity to grow food was limited by the amount of fertilizer available. The demand for ammonia was growing, but most ammonia had to be mined or collected from far-away guano deposits which were rapidly diminishing.[11] Chemists made several attempts to figure out a way to synthetically produce the compound, and in 1909 German chemists Fritz Haber and Carl Bosch developed a method that could pull nitrogen from the atmosphere and produce ammonia.

---

11   Hager 2008, 38.

It is hard to overestimate the impact of this discovery. The ability to mass produce fertilizer dramatically increased crop yields. The process was considered to make "bread from the air."[12] The discovery would earn Haber and Bosch a Nobel Prize and soon ammonia would be mass produced on an industrial scale. Today, it is estimated that 50% of nitrogen found in human tissue originated from the Haber-Bosch process.[13] It was also partially responsible for a population explosion since the planet could now sustain larger crop yields. Since 1900 the global population has grown from 1.6 billion to 7.9 billion.[14]

It might be tempting to think that Haber's efforts to produce ammonia did a lot of good for the world and that this suggests once again the benefits that pure science offers to society. However, Haber's research took a dramatic turn after 1914 when Germany and the Entente Powers declared war. Scientists were recruited by national governments to help in the war effort. For example, with Germany being blockaded its ability to obtain guano to produce the nitrates required for explosives was limited, but the Haber-Bosch process allowed Germany to produce the explosives on a mass scale and prolonged the war by years.[15] The same process that could help feed billions also helped produce weapons of mass destruction.

Moreover, Haber joined ninety-two other German intellectuals in signing the *Manifesto of the Ninety-Three*, a document composed by scientists, scholars, and artists declaring their unequivocal support for German military action during World War I. Haber declared, "During peace time a scientist belongs to the world, but during war time he belongs to his Country."[16] Haber worked for the German Ministry of War where he developed a process whereby liquid chlorine could be converted into a gas and then projected across a battlefield into enemy trenches. His efforts to develop the means to deploy and weaponize chlorine gas would earn him the notorious distinction of being the "father of chemical warfare."[17]

---

12   Dekkers 2018, 179.
13   Ritter 2008.
14   Smil 1999, 415.
15   Hager 2008, 168.
16   Herrlich 2013, 760.
17   Charles 2005.

Chlorine gas causes inflammation in the eyes, nose, throat, and lungs. In large quantities it can cause the body to produce so much mucus that it can cause asphyxiation. The manufacture of chemical dyes for textiles by the German chemical industry also produced chlorine gas as a by-product, making it cheap and easy to source.[18] On April 22, 1915, Haber was on hand during the Second Battle of Ypres as 168 tons of chlorine were deployed against French, Belgian, British, and Canadian troops. Those that didn't die were forced to flee in terror or urinate on a handkerchief and breathe through it.[19] The act was a violation of the Hague Convention of 1907 which prohibited the use of poisonous gas, but Haber defended the use of chemical warfare, saying, "Gas as a weapon is not in the least more cruel than flying pieces of metal."[20]

The example of Fritz Haber and the experience of chemists and other scientists during World War I offer insights into how we can practically understand the relationship between pure and applied science and how we think about the moral responsibilities of scientists. Haber's process for producing synthetic ammonia was certainly a significant discovery, but the very process that made fertilizer and produced "bread from the air" could be used to make artillery shells and bring death from the air. As Haber would explain, "The development of the chemical industry made the effective renewal of this form of combat obvious."[21]

It might be easy to say that there is a fine line in theory between the science that leads to discoveries and the science that applies them, but in practice things have a way of bleeding over and often for very specific reasons. The same science that enabled the industrial revolution enabled industrial warfare. Without mass production and industrial mobilization, warfare on the scale of World War I would not have been possible. The fact that it can be so difficult to anticipate the long-term ethical consequences of a scientific discovery is part of the reason it may not be wise to make such a distinction between pure and applied science in the first place. When we reflect on the Wiley example, we might consider the potential future human cost of developing certain kinds of research. Can we also imagine a data scientist or AI developer in the

---

18   Hager 2008, 84.
19   Edmonds and Wynne 1927 [1995].
20   Haber 1920.
21   Haber 1923.

not-too-distant-future echoing Haber's sentiments that in wartime they belong to their country rather than the world or humanity?

Following the experiences of World War I, many reconsidered the relationship between pure science and applied science, and whether scientists should be morally responsible for any harms that their discoveries permit. And as the interwar period led to World War II and eventually the Cold War, the debate would only become more paramount. Thus, to answer our questions about scientists' responsibility, our inquiry will now consider this debate.

3. The Social Purposes of Science and Scientific Responsibility

In the interwar period, just as many were beginning to seriously consider the relationship between science and society and the moral responsibilities of scientists, the wheels were already in motion to translate the latest discoveries in physics into the first nuclear weapons. While the war would create some disillusionment with science, many continued to advocate science and its benefits in the 1920s as the rise of mass production created cheaper prices and greater profits, and inventions like the radio and the vacuum cleaner made life more convenient at home.[22]

However, by the 1930s and the Great Depression, the mindset of the public and of scientists themselves began to shift. Not only did it look like the world might be on the path to war where once again science would magnify human suffering, but mass production had a profound effect on the prices of goods and labor, causing lower wages and higher unemployment. With the popular sentiment that individual wage earners were paying the cost of scientific progress, there were even calls for moratoriums on further research, just as some now call for a pause on AI research. The misery of the depression caused many to question the notion that scientific progress and human betterment went hand in hand.

While some still defended science by distinguishing between pure and applied research, it became difficult to defend pure science because during the previous decade, scientists sought funding for pure research

---

22   Kuznick 1989.

on the basis of its eventual applications. Thus, "by mid-decade, even the scientists themselves were beginning to tire of the facile demurs of those who insisted that they were simply adding new knowledge that was malevolently employed by others or that the beneficial uses of science greatly outweighed the harmful ones." However, "a new consciousness of the social responsibility of science and its practitioners swept the nation in the late 1930s."[23]

In 1939, Irish Scientist J.D. Bernal published a book titled *The Social Function of Science*, which opens as follows:

> The events of the past few years have led to a critical examination of the function of science in society. It used to be believed that the results of scientific investigation would lead to continuous progressive improvements in conditions of life; but first the War and then the economic crisis have shown that science can be used as easily for destructive and wasteful purposes.[24]

Bernal's study found that while science may claim to seek objective truths for its own sake, as science moved into the twentieth century, it was largely being funded by state and financial interests, and the application of science was mostly applied to unlocking new sources of material profit, greater manufacturing efficiency, and new capacities for war making.[25] Science has the potential to serve many social purposes, but far from the notion that science naturally benefits society, science is held captive by the vested interests that fund it.

Philosophers have also argued that the distinction between pure and applied science is itself ethically problematical. John Dewey wrote:

> It is an incident of human history, and a rather appalling incident, that applied science has been so largely made an equivalent of use for private and economic class purposes and privileges. When inquiry is narrowed by such motivation or interest, the consequence is in so far disastrous both to science and to human life ... It springs from defects and perversions of morality as that is embodied in institutions and their effects upon

---

23   Kuznick 1989, 38.
24   Bernal 1939, xiii.
25   Bernal 1939, 10.

personal disposition. It may be questioned whether the notion that science is pure in the sense of being concerned exclusively with a realm of objects detached from human concerns has not conspired to reinforce this moral deficiency.[26]

Many began to take note of the relationship between science and industry, the dependence of science on industry, and to see scientists as defenders of an unequal social order.

For example, in the 1920s, roughly 70% of chemists worked in industry.[27] Many scientists internalized the worldview of their business and industrial allies. The American Chemical Society not only justified the involvement of chemists in chemical warfare research but lobbied against the Geneva Protocol designed to outlaw such weapons. In response, sociologist Read Bain's 1933 article "Scientist as Citizen" was scathing. He wrote, "Scientists, with few notable exceptions, are the worst citizens of the Republic ... they sneer at politics and politicians ... They sell their services to exploiters of human life ... They produce powerful mechanisms and proudly proclaim that they 'do not care how they are used—leave that to the moralists,' ... The 'pure' scientist has to be a moral eunuch or a civic hermit."[28]

With shifting perspectives on the role of the scientist in society, there were calls for scientists to be more proactive about the application of science for the benefit of humanity. A 1937 report on a meeting of the American Association for the Advancement of Science noted "the increasing sense of responsibility of scientist to society. As an editorial in the Washington Post expressed it, 'the current movement might be described as an effort to shift from science for science's sake to science for the sake of humanity.'"[29] Princeton biologist Edwin G. Conklin argued that "The spirit of science and the method of science must spread to society and government."[30] MIT mathematician Leonard M. Passano urged scientists to consider the social effects of their research, especially when the "chief aim of scientific research is to enable those who already

---

26   Dewey 1929, 164–65.
27   Kuznick 1989, 10.
28   Bain 1933, 413–14.
29   Moulton 1938, 95–96.
30   Kuznick 1989, 63.

receive an undue share of wealth produced by industry and research, to appropriate a share larger still."[31]

However, Bernal also claimed that it isn't up to scientists to decide for themselves what their roles and responsibilities in society are. "The question ... of whether science can be recognized at all is not simply or even principally one for scientists," he argued. "It is a social and political question."[32] But such a reorganization of science for the common good might mean banning or limiting research: the freedom of the scientist being curtailed by political interests. This challenges the linear model and its definition of scientific freedom.

3.1 OPPENHEIMER AND THE NUCLEAR BOMB

Ironically, as these debates were taking place, science was marching towards the dawn of the nuclear age. Before the beginning of the twentieth century, Pierre and Marie Curie had discovered radium and radioactive decay. Shortly afterwards, Albert Einstein published his theory of special relativity, which suggested that splitting atomic matter could yield massive amounts of energy. By 1938, Otto Hahn, Lise Meitner, and Fritz Strassmann had managed to bombard uranium with neutrons and to discover nuclear fission. Much of this research was driven by a desire to learn more about the physical world, and while some believed that little use would come of this ability, it also didn't take very long for people to realize the potential applications. Just five years earlier, a Hungarian physicist named Leó Szilárd patented his idea to create a self-sustaining nuclear chain reaction which would allow for massive amounts of energy to be released, and began to worry about the potential to apply these ideas.[33]

After the rise of Hitler and the Nazi regime in 1933, Szilárd fled to England. When he heard about the discovery of fission in 1938, he realized that uranium could be used in a chain reaction. Worried about the potential applications for making a bomb and feeling a sense of moral responsibility, Szilárd reached out to Einstein to warn him about the possibility of nuclear weapons. Einstein had never considered that

---

31   Passano 1935, 46.
32   Bernal 1939, 241.
33   Lanouette and Silard 1992, 102.

possibility.[34] They wrote a letter to American President Franklin D. Roosevelt warning him about the risks: "It may become possible to set up a nuclear chain reaction ... by which vast amounts of power ... would be generated. It now appears almost certain that this could be achieved in the immediate future. This new phenomenon would also lead to the construction of bombs," the letter warned.[35]

Roosevelt responded by creating the Manhattan Project and developing the first nuclear bomb. Spearheaded by physicists like Robert Oppenheimer, the project led to the first testing of an atomic bomb on July 16, 1945. On August 6, the first atomic bomb was dropped on Hiroshima, followed by a second bomb on Nagasaki three days later. Japan surrendered on August 15 and the war was over. It is estimated that over 200,000 people died in the bombings, with countless suffering from wounds and radiation related illnesses. Five years later, following the efforts of spies on the Manhattan Project, the Soviet Union conducted its first nuclear bomb test and for the next five decades the world would be gripped by the threat of nuclear war and global destruction.

The scientific world's reaction to the development and use of nuclear weapons was mixed. Many who worked for the Manhattan Project later expressed regret for participating.[36] Oppenheimer reported that as he observed the first nuclear test, a passage from the *Bhagavad Gita* came to his mind: "Now I am become Death, the destroyer of worlds." Unlike Fritz Haber, Oppenheimer regretted the suffering he had caused and moreover worried about the potential future suffering his work could cause humanity. He later reported to President Truman, "Mr. President, I feel I do have blood on my hands."[37]

See Oppenheimer discuss his creation.

In the span of a few decades, just as scientists were beginning to consider their social responsibilities, science laid the groundwork for advancements in nuclear technology that would change the entire world. Who was responsible? Are Marie and Pierre Curie responsible for nuclear weapons? What about Hahn, Meitner, and Strassmann? Without

---

34   Lanouette and Silard 1992, 199.
35   Einstein 1939.
36   Ham 2015.
37   Hunner 2012, 151.

the discovery of nuclear fission, atomic weapons would not be possible. What about Szilárd and Einstein? Their letter warned Roosevelt that the Germans could be developing their own atomic weapons, but in hindsight the Germans weren't remotely close to building nuclear weapons.[38] Had they not sent their letter, the Americans might not have built and used an atomic bomb. Einstein regretted his role in the letter for this very reason, telling *Newsweek* in 1947, "had I known that the Germans would not succeed in developing an atomic bomb, I would have done nothing."[39] How do we determine responsibility in these cases?

4. Bridgman's Defence of Scientific Freedom

One defender of scientific freedom following World War II was physicist Percy Bridgman. Bridgman's defence of the distinction between pure and applied science and the importance of discovering scientific truth are notable, but in context of the late 1940s his argument wasn't just that scientists should be free to pursue truth because science is important, but that it is wrong to expect scientists to be responsible for the consequences of their research. His 1948 article "Scientists and Social Responsibility" highlights some important issues that we must consider if we are to better understand the kind of responsibilities that scientists might have for AI development today. He charges that the imposition of scientific responsibility would force scientists to engage in work for which they are ill-suited, that such a move is exploitative, and that ultimately it is the job of society itself to manage scientific discoveries.

Bridgman begins by asking what scientific responsibility means. What does it mean to say that science is responsible for the atomic bomb or the Great Depression? Many would say that scientists are responsible for their discoveries. One way of understanding such a claim is to say that science is causally responsible. The discovery of fission enabled the construction of a nuclear bomb, so scientists are causally responsible for the development of nuclear weapons. But there is a difference between **causal responsibility** and **moral responsibility**. For example, if I save someone's life by accident, I might be causally responsible for

---

38   Walker 1995, 198–99.
39   *Newsweek* 1947.

doing so, but it would be harder to argue that I am morally responsible or deserving of much praise. In other words, when we are concerned about moral responsibility, we are concerned with attitudes of moral praise and blame.

Fritz Haber is causally responsible for the development of synthetic fertilizer, but is he morally responsible for the population boom that followed? Is he morally responsible for all the effects of this, including the impact on the climate? It seems difficult to say that he is morally responsible for all those things, but is Haber morally responsible for the development and deployment of chemical warfare? It's easier to say that he is in this case because he directly intended to create chemical weapons to win the war. But perhaps it is simply easier to see it this way because Haber was far more directly causally responsible for the development of chemical weapons than for global population trends a century after the fact. Where moral responsibility takes over, what the scope of the responsibility is and where it comes from remains unclear.

We must consider who exactly has personal responsibility. To say that science is responsible for its discoveries presumably means that scientists themselves are responsible. But what does that mean? Bridgman understands the claim to mean that "each and every scientist has a moral obligation to see to it that the uses society makes of scientific discoveries are beneficent." However, he argues that this would be morally wrong, as "the discipline that would be imposed is the ... loss of scientific freedom."[40]

First, let's consider whether it would make sense to say that all scientists do share the same responsibility. To start with, not every scientist can always be aware of how their work may be used. Moral responsibility isn't usually extended to all consequences that follow from a discovery. As Bridgman argues, "The miner of iron ore is not expected to see to it that none of the scrap iron which may eventually result from his labors is sold to the Japanese to be used against his country."[41]

Also, Bridgman believes that it would be wrong to expect all scientists to be concerned about the effects of their work. It would be too much of a limitation on the scientist's time and freedom, as he explains: "If I personally had to see to it that only beneficent uses were made of

---

40   Bridgman 1948, 69.
41   Bridgman 1948, 69.

my discoveries, I should have to spend my life oscillating between some kind of forecasting bureau, to find what might be the uses made of my discoveries, and lobbying ... to procure the passage of special legislation to control the uses."[42] This would be too much of an imposition on the individual scientists and forces them to focus on social and political issues rather than specialize in their field. Consider again the discussion in Chapter I about the logistical difficulties involved with establishing an ethics review board for AI.

If scientists are expected to be socially responsible, Bridgman argues that this should be its own field of research, one that could attract specific scientists who are interested in scientific responsibility rather than be the concern of each scientist. Moreover, if responsibility is going to be imposed on scientists, scientists need support from society to exercise that responsibility.

Instead, Bridgman argues that regulating the discoveries of scientists is the job of society itself. As he notes, "The applications of scientific discoveries are very seldom made by scientists themselves, but are usually made by industrialists. It is the manufacture and sale of the invention that should be controlled rather than the act of inventing."[43] For example, when it comes to nuclear weapons, he argues, "If [society] had not wanted to construct an atomic bomb, it need not have signed the check for two billion dollars which alone made it possible. Without this essential contribution from society the atomic bomb would have remained an interesting blueprint in a laboratory."

As noted, Bridgman defends the pursuit of scientific truth for its own sake, and so scientists must remain free to discover things about nature. The application of this knowledge is not a concern of the scientist and so he urges, "let the scientists, for their part, take a long-range point of view and not accept the careless imposition of responsibility, an acceptance which to my mind smacks too much of appeasement and lack of self-respect."

Bridgman's argument makes some good points, but it also contains several questionable assertions. First, it would require additional time and resources for scientists to be active in the regulation of the applications of their discoveries. Also, once a discovery is made it is often

42   Bridgman 1948, 70.
43   Bridgman 1948, 71.

beyond individual scientists' control how that discovery will be used, meaning that if we expect scientists to be responsible for their work, then society must listen to scientists when they warn of societal dangers.

Secondly, it is difficult to say exactly who should be responsible for what when it comes to science. As we've learned from the history of the development of nuclear weapons, scientific breakthroughs often involve a long chronological chain of smaller discoveries with later scientists building on the work of earlier ones. It may be easy to say that Oppenheimer bears some responsibility for the development of nuclear weapons, but how responsible is Marie Curie? Her discoveries in radioactivity helped create nuclear weapons, but they weren't something that she anticipated. If an invention couldn't be anticipated, does that mean that the discoverer of the process behind it bears no responsibility, or should scientists still be held responsible for not unleashing discoveries on the world when they can't anticipate how the world will react?

At least we can say that it is unfair to expect scientists to foresee the future. At the same time, Bridgman oversimplifies many of the issues he discusses. For example, he notes that most applications of science are made by industrialists rather than scientists. However, the actual lived history, as we've discovered, suggests a far cozier relationship than Bridgman would care to admit. In the 1920s and 1930s, chemists were far more likely to embrace conservative causes, largely owing to their close ties to industry and their participation in chemical warfare research.[44] In the 1920s roughly 70% of all chemists in America worked in industry and were paid significantly higher wages than other scientists.

The notion that scientists have no idea what industrial and chemical warfare research is going to be used for, or that scientists are not especially rewarded for their efforts specifically because of this fact, stretches credulity. Sometimes it is obvious what a scientific discovery could be used for, just as it was obvious to Szilárd and Haber. The two billion dollars provided for the Manhattan Project only came after the urgent prodding of the Szilárd-Einstein letter warning that other scientists could be building their own bomb. In other words, Bridgman simplifies matters too much by suggesting that scientists are not suited to lobbying for regulation by ignoring the lived experience of scientists themselves.

---

44 Kuznick 1989, 61.

It can also be better to prevent a problem altogether by simply not engaging in certain kinds of potentially controversial research at all rather than by trying to fix the problem with regulation after the fact. In many cases once the cat is out of the bag it is difficult to do much about it. Consider ethnic facial recognition, for example. We can imagine how much harm such a technology could cause, but trying to regulate its uses will be more difficult than if it were simply not developed at all. Also, many data scientists and AI developers are highly sought after and well rewarded for their efforts, and so the question remains how conscientious they should actually be about how their research will be used.

Lastly, Bridgman seems to have fixed and outdated notions about the job of being a scientist. By the twentieth century the role of science in society had massively changed, both in terms of the degree of proliferation of the use of science in society but also in the amount of public investment in scientific education, usually with large and structured funding packages paid with taxpayer money. With more scientists now pursuing more scientific research, the job of scientists and its role in society has clearly changed. However, Bridgman's notion that responsibility must be imposed from without and that to accept it would be undignified for scientists seems to preclude the idea that the role of scientist would change as science gains prestige and influence in society.

After all, we expect lawyers, accountants, or medical practitioners to conform to professional codes of ethics, or we pass new legislation regulating such fields, because we recognize the increasing importance of such jobs and their potential for abuse. So, why should the job of science be excluded from similar developments, particularly in the face of changing social circumstances? This point also highlights the problematic issue of "proportionality" that Bridgman refers to. His argument hinges on the notion that if society holds scientists responsible for their research, it will exact disproportionate service from them. But what exactly is a proportionate level of service? If we introduce new regulations for professional accountants to improve transparency and accountability, is this an example of this society exacting disproportionate service for the job or simply an evolution of the job itself?

The role of the scientist will always be determined by who is investing the money for it and whatever laws regulate the field. A society, particularly a democratic one, must be able to govern its own affairs and scientists are free to practice under those conditions or not. Thus, proportionality of service is going to be an evolving concept. There may be questions about where certain restrictions on research would constitute a violation of the scientist's civil rights, but there must be give and take. The scientists' right to discover and publish research on nuclear weapons doesn't automatically override citizens' rights not to be killed in a nuclear explosion.

Indeed, since Bridgman's time scientists and philosophers have debated the problem of **forbidden knowledge**. Some of the worst abuses in science happened in the name of experimentation, and currently there are plenty of laws and regulations that prevent certain kinds of research if they involve experimentation, particularly on people. But the topic of forbidden knowledge also includes areas of study that some believe should be off limits for scientific research. Human cloning is a good example of the kind of research that has been banned.

However, there are more nuanced forms of research that some claim should be forbidden as well. In fact, philosopher Janet Kourany argues that because science is funded by society through the form of taxes and consumer spending, scientists have moral responsibilities to act according to egalitarian standards. If a scientist wants to study any differences between racial and gender groups, they should seek to explain them as far as empirically possible without assuming that the difference is biologically determined.[45] According to Bridgman, this would likely seem like too much of an imposition on the scientist.

But there is one final point to consider. In a response to Bridgman's article, chemist Harold C. Urey writes, "I believe that responsibility in our society is voluntarily accepted by people, and not imposed on them, and hence that Professor Bridgman's whole argument in regard to the imposition of responsibility is beside the point."[46] Indeed, Bridgman's argument seems to focus on whether responsibility would be a limitation of freedom, not whether scientists as a matter of ethics should take responsibility on their own. In fact, it isn't all that clear whether some of the responsibilities we have been discussing are not imposed on the scientist as a scientist, or whether we all naturally should take responsibility for certain kinds of actions. This is especially important, as we

45   Kourany 2010, 72.
46   Urey 1948, 72.

shall see, because we are all held accountable for certain kinds of actions, especially for not being reckless or negligent.

5. Negligence and Recklessness

Bridgman's argument supports the notion that applied science and pure science are distinct and that when engaged in pure science, a scientist should keep value judgements out of their research. But in 1953 philosopher Richard Rudner published an article titled "The Scientist Qua Scientist Makes Value Judgments." He argues that scientists make value judgements when it comes to deciding whether to accept a hypothesis. He explains: "no scientific hypothesis is ever completely verified"; thus, "in accepting a hypothesis the scientist must make the decision that the evidence is sufficiently strong or that the probability is sufficiently high to warrant the acceptance of the hypothesis."[47]

Imagine that a scientist tests a drug for safety and knows there's a chance that their test could be wrong. Even if the test shows evidence that the drug is safe, there's still a chance that the test is wrong. How sure should you be before declaring that the drug is safe? As Rudner explains, "our decision regarding the evidence and respecting how strong is 'strong enough', is going to be a function of the importance, in the typically ethical sense, of making a mistake in accepting or rejecting a hypothesis ... How sure we need to be before we accept a hypothesis will depend on how serious a mistake would be."[48] This is an example of **inductive risk**, or the risk of error in accepting or rejecting a hypothesis. Since it is the scientist who makes this decision, and since this decision is ethical in scope, then clearly the scientist must be held responsible for these value judgements.

But why should scientists care about inductive risk? Can't they simply ignore such concerns? According to philosopher of science Heather Douglas, the answer is no. Inductive risk considerations are questions about the consequences of error and they stem from considerations about recklessness and negligence. According to Douglas, individuals should be held ethically accountable for not being reckless or negligent

---

47   Rudner 1953, 2

48   Rudner 1953, 2.

unless there is a reason for a special moral exemption. She argues that scientists do not deserve moral exemptions from being reckless or negligent and thus are as responsible as the rest of us for considering the consequences of their errors.

Moral responsibility extends to things we intend to do, but also certain actions we don't intend. If a driver doesn't regularly inspect their vehicle and its wheel comes off and causes an accident, we say that the driver is morally responsible even if they did not intend that result. **Negligence** occurs when a person takes an action not knowing that they are risking harm, but they should be aware of the risk. Alternatively, **recklessness** occurs when someone knows the risks of harm that their actions could cause but chooses to do them anyway.[49]

In the early twentieth century, mechanical engineer Thomas Midgley Jr. developed a solution to a problem that plagued many early automobiles called "knocking" by adding a lead-based compound called tetraethyllead to gasoline. Midgley worked for General Motors and knew that his discovery would make a lot of money. The problem is that lead was well known, even as early as 1920, for its toxic effects, causing hallucinations, insanity, and death.[50] Despite this, and although workers at the factories producing the fuel (and even Midgley himself) became sick with lead poisoning, they proceeded to sell it and would insist on the safety of leaded gasoline for decades.

Eventually some workers began to die. In 1924 at a press conference Midgley attempted to demonstrate how "safe" the gasoline was by inhaling its vapors for a whole minute.[51] Midgley, who knew the risks, always advised the public that it was safe and recklessly insisted on the product. For decades lead was burned in gasoline and released into the atmosphere, until leaded gasoline was phased out in the 1970s. However, as a result of the amount of lead released into the atmosphere, everyone was exposed to higher concentrations of lead, resulting in higher crime rates, deaths, and a general decline in human IQ levels.[52]

Midgley's other major contribution was the development of Freon gas, which is used as a refrigerant. Prior refrigerants were flammable or toxic, but Freon is nontoxic and was eventually used in refrigerators and

---

49   Douglas 2010, 68–72.
50   Markowitz and Rosner 2013, 43.
51   Markowitz and Rosner 2013, 22.
52   McFarland, Hauer, and Reuben 2022.

in aerosol products. The problem is that Freon is a kind of chlorofluo-rocarbon (CFC) and it was later discovered that this gas goes into the atmosphere, where it remains and occasionally releases chlorine atoms which react with ozone, breaking it apart. Midgley's invention and other CFCs were discovered to be destroying Earth's protective ozone layer. CFCs were eventually phased out as well. In this case we might say that a reasonable person should have been aware of the major side effects of such a widely used product, and that because Midgley did not test this first, he was negligent in the development of Freon. For these two reasons, environmental historian J.R. McNeill has said that Midgley "had the most adverse impact on the atmosphere than any other single organism in Earth's history."[53]

These two cases demonstrate how a scientist can be reckless or negligent and what can be the consequences. One benefit of understanding moral responsibility in terms of recklessness or negligence is that these concepts do not rely on the notion of perfect foreknowledge of the future. No one expects the negligent driver to predict the future, but certain consequences are reasonably foreseeable, and in these cases we expect the person to do due diligence and consider the consequences that we could reasonably predict.

Bridgman might reply that ultimately the choice to accept a hypothesis is a matter of applied science, a choice for industrialists or policy makers. Scientists could advise whether something is likely or not likely to be the case, but the decision to proceed is not the scientists'; hence they are not responsible for what happens. And to maintain the integrity of pure research, scientists should be exempted from such ethical considerations.

Douglas believes this argument doesn't work. In some cases, it is acceptable to exempt certain people from certain kinds of moral responsibilities because of the role they are fulfilling. For example, a defence lawyer who is aware of their client's criminal wrongdoing is not obligated to report it because of lawyer-client confidentiality.[54] Douglas argues that scientists shouldn't be permitted similar exemptions.

To begin with, as a society we don't generally place such a value on scientific truth that we would pursue it no matter the cost. We would have

---

53   McNeill 2001, 421.
54   Douglas 2010, 73.

to accept that scientific truth is more important than all other ethical concerns, which we generally don't. For example, most forms of human experimentation are banned because we value human rights more than scientific truth. Secondly, role responsibilities typically require a system (like the criminal justice system) where moral responsibilities can be clearly defined, and others can shoulder someone else's moral responsibility. In the previous chapter we discussed how, for example, physicians can consult ethics review boards to assist them in complicated ethical cases. But such a system would not work for science in general.

Douglas argues that this is because science usually deals with uncharted territory, where only the scientists themselves can "*fully* appreciate the implications*" of their research.[55] Even Bridgman seems to acknowledge this, noting that "there are certain aspects of the relation between science and society that the scientist can appreciate better than anyone else, and if he does not insist on their significance no one else will."[56] Also, if scientists did not bear the burden for the consequences of error in their research, they would need constant ethical oversight by others, which is not only impractical but also not likely wanted by scientists themselves. Ethical review boards would have to constantly monitor the scientist, limiting their autonomy.

So, there is no good reason for scientists being freed from the general moral responsibilities not to be reckless or negligent for the sake of truth; furthermore, there is no one else that could shoulder those responsibilities. Scientists have the same moral responsibility as the rest of us to at least consider the consequences of error in their research and to be aware of the potential side effects of their research. As we will discuss further in Chapter 4, the issues of recklessness, negligence, and inductive risk are important concepts in the development of AI as we consider the consequences that may occur if an artificial intelligence gets it wrong.

INQUIRER'S TOOLBOX

6. When I consider a moral solution as an end-in-view, does it make sense? Is it a practical workable solution?

---

55   Douglas 2010, 73.
56   Bridgman 1948, 71.

6. Moral Responsibility and AI Development

Modern science is not that old relatively speaking and the debate about the moral responsibilities of science and scientists has been evolving. Yet, our discussion about the origins of the concept of scientific responsibility can inform our understanding of the ethical development of AI. As we've discussed in the Wiley case, it can be difficult to maintain a clear moral distinction between pure and applied science, meaning that there is a legitimate question about whether scientists and other AI developers should be held morally responsible for their research, even if that research is only focussed on developing AI itself. Given that science has a profound impact on society and is funded by it, many argue that science has a social responsibility to not make our lives worse off for it.

Should AI developers be exempt from moral responsibilities? We may wish to distinguish between the development of facial recognition and its potential application towards the oppression of ethnic minorities by saying that the scientists only developed the algorithm but didn't put it into practice. But what about errors? An algorithm might erroneously label someone as belonging to an ethnic minority and because of this error they could encounter oppression. It is no longer so easy to make that distinction in this case because the scientist is responsible for the error.

Some AI scientists and developers who use AI for their products have recognized their ethical responsibilities. In 2023 the so-called "Godfather of AI" Geoffrey Hinton quit his job at Google to speak out more openly about the risks of AI destabilizing society and exacerbating income inequality. "I console myself with the normal excuse: If I hadn't done it, somebody else would have," he told the *New York Times*.[57] Reflecting on the part he played, former Facebook executive and engineer Chamath Palihapitiya said, "I feel tremendous guilt ... I think we have created tools that are ripping apart the social fabric of how society works."[58] His remarks related to the addictive and destructive qualities he perceives in Facebook's algorithms. Indeed, in discussing his role in using AI to exploit vulnerabilities in human psychology for Facebook, former

---

57　Metz 2023.
58　Vincent 2017.

president Sean Parker noted that they "understood this consciously—and we did it anyway."[59]

In 2021, Facebook data engineer Frances Haugen came forward as a whistleblower, informing the public that Facebook knowingly develops its algorithms to exploit the insecurities of young people, and that content on their site promotes violence.[60] Timnit Gebru, a computer scientist who worked for Google on their Ethical Artificial Intelligence Team, left Google in 2020 over her concerns about the ethical risks of large language models and has since been an advocate for several ethical issues in AI, including bias. The fact that corporations are consciously aware of the risks of their services but choose to proceed nonetheless is a recklessness that scientists, developers, and employees are morally responsible for. Just as in the 1930s, the debate surrounding the moral responsibilities of scientists and engineers is not going away, particularly when it comes to the wisdom of proceeding with research when you don't know what the outcome will be.

Nor is the discussion regarding forbidden knowledge and moratoriums on further development of AI going to go away. In 2023, an open letter signed by over 1,000 AI researchers called for a six-month pause on developing AI more powerful than the current iteration of ChatGPT because of ethical concerns.[61] As it explains, "Powerful AI systems should be developed only once we are confident that their effects will be positive and their risks will be manageable."[62] Some have even called for more extreme restrictions given the potential risks. Eliezer Yudkowsky, a decision theorist, points to the potential risks that AI could pose to life on Earth and calls for ending all development of powerful AI. Yudkowsky believes that an international treaty should ban its development with military enforcement, noting that we should be "be less scared of a shooting conflict between nations than of the moratorium being violated; be willing to destroy a rogue datacenter by airstrike."[63]

Others argue that moratoriums are not practical. Over 400 billion dollars was invested in AI in 2022 alone. Convincing people to give up this research, and all the potential moral and monetary benefits it might

---

59   Raicu 2020.
60   Allyn 2021.
61   Goldman 2023.
62   Future of Life Institute 2023.
63   Yudkowsky 2023.

bring, particularly if the risks of something going wrong are unclear, will not be easy. Even if we attempt to ban some forms of AI development, as Hinton noted, someone else will attempt to develop it. Still, this does not exempt researchers and developers from their responsibility and the choice they make to ultimately participate. If a scientist or AI developer assists in the development of AI, they are responsible for considering whether their actions are reckless or negligent and they are responsible for inductive risk. We will consider these moral responsibilities further in Chapters 4, 5, and 6.

7. Conclusion

One major point of contention that arises from the topic of scientific responsibility is whether there is a meaningful distinction between pure science and applied science. But for our purposes, the question is not necessarily whether we can distinguish between pure science and applied science, but whether this is an ethically relevant distinction. As we've seen from the historical record, in practice, such hard distinctions between pure and applied science do not always exist. As we can see in the case of Szilárd, what seems like useless science can very quickly transfer over to applied science in sudden, dangerous, and unpredictable ways. In many cases this goes beyond merely whether we want to praise or blame an individual scientist for their efforts, but rather acknowledging that science has changed the world in ethically complicated ways that we can't reverse. So, perhaps practically and ethically, the distinction doesn't really matter at all.

Another issue we might consider is the ethical importance of scientific freedom. As we've learned from Douglas, we don't always prioritize scientific truth over all other ethical concerns. We must balance the relative ethical importance of scientific freedom compared to other interests. When we consider scientific responsibility, should this take the form of laws or regulations, or funding packages where society directs scientists what to study or not study? Or should it take the form of voluntary restraints on research by scientists acting according to their own moral consciousness?

As we proceed through each chapter, we will encounter numerous examples of AI development that could potentially have serious ethical consequences. Are the people who develop addictive algorithms for social media morally responsible for its addictive qualities? Does a researcher have a duty to ensure that racist or sexist assumptions aren't underlying their models, and are they responsible for the harms those assumptions cause in the real world? The next chapter considers some of these issues in more depth as we investigate how biases can be embedded in AI. We will later consider what is morally at stake when AIs are in error.

ADDITIONAL MATERIAL

Pure science • science that investigates the world purely for the sake of knowledge.

Applied science • science that is applied to the world to achieve a goal or end.

Linear model of scientific development • investment in pure science will eventually aid in the development of technological and economic innovation through applied research and development.

Causal responsibility • cases where someone causes something to happen, but we don't hold them ethically accountable for causing it.

Moral responsibility • cases where someone causes something to happen, and they are held morally accountable for their actions. We would praise or blame someone for what they do.

Forbidden knowledge • areas of study that some argue should be off limits for scientific research.

Inductive risk • the risk involved in accepting or rejecting a hypothesis in error, based on the consequences of that error.

---

1. Should scientists be held responsible for the moral consequences of their work?

2. Do scientists and engineers owe a duty to their country in wartime or are they obligated to act from an international perspective?

3. Can scientists really ignore the potential applications of their research and leave it to the rest of society?

4. Are there some topics in science that we shouldn't pursue solely for ethical reasons?

5. Should Wiley have retracted the article? Does the publisher or do the researchers in the ethnicity recognition case have any ethical responsibilities for how that research might be used?

6. Given his accomplishments in aiding in food scarcity and in developing chemical warfare, how should we evaluate Fritz Haber?

7. How should society take advantage of the benefits of science? Should science serve private interests, or should science respond to democratic interests?

8. Given that AI development can include data scientists, computer engineers, business investors and managers, and many other people, how responsible is each of them for the consequences of their product? Are they all responsible for not acting recklessly or negligently?

9. Where is the dividing line between pure and applied science? Did studies into the requirements necessary to create a nuclear chain reaction constitute pure science or applied science?

10. Do you agree with Bridgman's view that it is the manufacture and sale of a discovery that should be regulated rather than discovery itself? Why or why not?

11. Is Fritz Haber as responsible for climate change as he is of chemical weapons?

12. In the chapter, we noted Read Bain's claim that "Scientists, with few notable exceptions, are the worst citizens of the Republic ... they sneer at politics and politicians ... They sell their services to exploiters of human life ... They produce powerful mechanisms and proudly proclaim that they 'do not care how they are used—leave that to the moralists.'" Is this a fair assessment of contemporary scientists who work in AI?

3

Bias and Machine Learning

In democratic societies under the rule of law, we expect our justice systems to be fair. We expect citizens to be charged, tried, and sentenced according to available evidence and due process, rather than relying on arbitrary factors that do not bear on the crime. In a fair criminal justice system, for example, we shouldn't expect major differences in the outcomes of cases if the only differences are the defendant's race or ethnicity. Despite this, criminal justice systems have long been known to suffer from biases which allow arbitrary factors such as skin colour or race to skew its findings and results.

A study from the University of Michigan found that in federal cases, Black people are likely to receive a sentence 10% longer than those given to white people for the same crime.[1] Black people also make up a disproportionate number of prisoners in federal prisons relative to their share of the total population. They are more likely to be disadvantaged in terms of sentence length, while Latinos are more likely to be disadvantaged over the decision to incarcerate.[2] Given these biases in the criminal justice system, many have considered replacing some of the decision procedures with artificial intelligence to determine things like sentence length or rehabilitation measures, the operative assumption being that

---

1   Rehavi and Starr 2014, 1320.
2   Kansal 2005.

using an algorithm will prevent the interjection of human prejudices and "take the guesswork" out of things.

The thinking goes that if we can replace humans, who have personal biases, with AI, which does not share our biases, we can create a more objective and fair process for determining criminal sentences. The White House has supported the use of AI to reduce systematic bias.[3] Several US states have begun using predictive recidivism algorithms to help determine whether defendants are likely to re-offend, whether they should qualify for rehabilitation programs, whether they should be held in jail before trial, and how severe their sentences should be. The hope is that such algorithms will make the criminal justice system fairer, more consistent, and more efficient.

Yet, the reality has been different. Many risk assessment algorithms have been demonstrated to reflect human biases. For example, one algorithm known as Correctional Offender Management Profiling for Alternative Sanctions (COMPAS) has been accused of racial bias, with Black people found twice as likely as white people to be labelled a higher risk for recidivism.[4] Communities that have been disproportionately targeted by police are more likely to be considered high-risk offenders. Such biases in data can cause biases in the output of the algorithm, as we will discuss in detail. As a result, AI being used in criminal justice is helping to amplify and perpetuate the same biases.

But how can a machine be biased? Computers don't have personal or political preferences, so is it human error that creates biases or is it the data that the AI has access to? As we shall soon discover, both issues contribute towards the creation of biased artificial intelligence. This chapter will consider four central questions:

1. What causes AI to be biased?
2. What are the ethical consequences of bias in artificial intelligence?
3. What do these cases tell us about moral thinking?
4. What are some possible solutions for eliminating bias in AI or mitigating its effects?

---

3   Shueh 2016.
4   Angwin et al. 2016.

Much of the ethically problematic examples of biased artificial intelligence stems from the use of machine learning to derive models that allow for the prediction of certain outputs given certain input data. To unpack the ethical issues involved we will briefly consider what machine learning is and how it relates to modelling. We will also consider the kinds of long-term ethical consequences of these models and what this can tell us about moral thinking and moral responsibility. Later, we will consider several case studies involving bias and AI.

1. What Causes AI to Be Biased?

In science, bias can prevent us from reaching reliable conclusions about the world. However, when scientific bias is applied in social situations, it can create or exacerbate social biases which can have more profound ethical consequences. If an algorithm like COMPAS is biased, not only will we not generate reliable predictions about who is and isn't likely to reoffend, but that bias will disproportionately affect any community the algorithm is biased against. Thus, a scientific bias can affect real world fairness in society.

Fairness, and the perception of it, is morally important because it is difficult to get people to participate in a process that they know is unfair. If people know that a test is biased, for example, the public won't trust in the test. One central concept of fairness is **equality**. In other words, it is fair to treat people the same unless there is a relevant difference between them. If two people share a similar education and work background, they should have roughly the same chances of getting an accounting job without something arbitrary like their physical appearance being the difference that led to one being hired over the other.

Using Kant's categorical imperative, we can understand the moral problem with using arbitrary tests to treat people unequally. We could not will a maxim permitting the universal use of arbitrary factors to stand in judgement of us, because we would always insist on an exception for ourselves and because if everyone knew the process was unfair, they would not trust it. If treating people equally means that we do not let arbitrary tests influence how we treat them, then we ought not let bias unfairly cloud how we treat people. This can become a morally problematic form of discrimination if we practice this at a large scale.

As we discussed in the previous chapter, a principle like "fairness" does not share consistent understandings. For example, treating everyone equally, regardless of where people started in life or what natural talents they may have or lack, might be considered unfair. Instead, to ensure everyone has a fair opportunity of having success, rather than treat everyone the same, greater opportunities should be provided to those that need them more. This is a form of fairness known as **equity**. Now we can turn our attention to bias in AI.

The question "What causes AI to be biased?" is slightly unhelpful. Not all forms of bias are inherently ethically problematic. A **bias**, in its most neutral meaning, is simply a tendency for something to go in one direction rather than another, to favour certain conclusions rather than others, or to seek certain outcomes rather than others. Some biases are desirable and could even be considered morally appropriate. For example, it is appropriate to have a bias towards doctors with medical degrees over doctors without them.

Many will argue that affirmative action programs that create a bias towards hiring marginalized people are a morally acceptable form of bias since their purpose is to build equity. As Dewey notes, "bias for impartiality is as much a bias as is partisan prejudice though it is a radically different kind of bias."[5] Our primary concern is, thus, not what causes AI to be biased, but to ask, "Why does AI generate biases that reflect morally problematic values?"

1.1 INTENTIONAL DESIGN

There are at least three reasons why AI generates morally problematic biases. The first is that the AI was designed to be that way. For example, the business models of many financial, educational, and advertising companies are predatory in nature, seeking to identify specific kinds of people to prey upon. In *Weapons of Math Destruction*, Cathy O'Neil details the efforts of for-profit universities and high-interest payday loan companies using targeted advertising to identify specific groups that would be susceptible to their appeal.

Algorithms can allow a company to target specific people with advertisements. O'Neil explains: "If the program is predatory, it gauges their

---

5   Dewey 1998, 212.

weaknesses and vulnerabilities and pursues the most efficient path to exploit them."[6] These ads are designed to pinpoint people in need and sell them on false or overpriced promises. For example, a complaint by the State of California charged that Corinthian College targeted people who were "isolated," "impatient," had "low self-esteem," had "few people in their lives who care about them," or who were "stuck."[7] These are algorithms designed to have a bias towards identifying the most desperate people to target them with advertising. As O'Neil notes, "increasingly, the data-crunching machines are sifting through our data on their own, searching our habits and hopes, fears and desires."[8] If the AI is designed to target vulnerable people, it will be biased because it was designed to be that way.

1.2 BIASES EMBEDDED IN MODEL

A second reason AI can generate biased results is because the biases are embedded by the developers unintentionally or inadvertently. If an algorithm is designed to be biased, it's no surprise that it is. Alternatively, AI developers might unintentionally embed certain biases in an algorithm. They might not want to design a biased algorithm, yet the decisions they make may reflect these biases and will nevertheless have this effect. If biases creep into how we understand concepts and relationships, machine learning will only serve to magnify such biases at a large scale.

Machine learning (ML) is "a subset of artificial intelligence that uses algorithms and statistical models for computers to perform specific tasks without human interaction."[9] It requires that the algorithm must learn from and improve through experience.[10] Learning from experience requires data to build a model of the relationships between input values to accurately predict certain output values. In **supervised learning** the inputs and output values are labelled by humans, meaning that the machine learning algorithm must produce a function that can predict the correct outputs based on input values. By contrast,

6   O'Neil 2016, 77.
7   O'Neil 2016, 71.
8   O'Neil 2016, 75.
9   Nandi and Pal 2022, 1.
10   Mitchell 1997, 2.

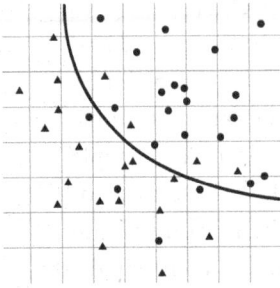

FIGURE 6 · Using statistical techniques like regression and classification, we can find relationships between changes in variables to discover patterns.

unsupervised learning involves no labeled data, and the learning algorithm must find the correct labels on its own.

Most forms of machine learning "learn" using statistical inference. They employ techniques like **regression** and classification to demonstrate that changes in one variable (**independent variables**) are associated with changes in other variables (**dependent variables**) (Figure 6). For example, if we wanted to know how much the amount of time studying will affect academic performance, we could collect a dataset containing time studied and the eventual student grades. Using regression, we could determine how much difference the amount of time spent studying would impact future grades. Perhaps we might find that studying ten hours a week has a significant impact on one's grades, but studying more than ten hours only has a marginal impact. We could include several additional variables and attempt to consider their impact on grades as well. Once a web of correlations between variables is detected, these relationships can be captured in an overall mathematical function that will accurately predict an output based on all the inputs (Figure 7). Machine Learning "works to uncover patterns in data, to build and refine representative mathematical models of data that can be used to make predictions and/or describe data to gain insight."[11] This is true whether our purpose is to detect recidivism based on criminal history or to detect handwriting based on the shade of a pixel.

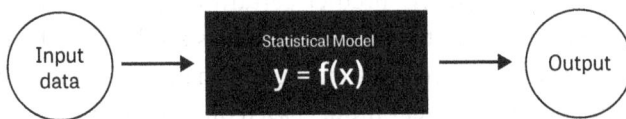

FIGURE 7 · The development of a model using machine learning aims to discover a mathematical function that will generate correct outputs (represented by y) given specific input data (represented by x). The solid line in Figure 6 represents the function.

11   Singh et al. 2016.

For our purposes, the important thing to note is the relationship between machine learning and modelling. Many AI-powered algorithms we'll consider can be more accurately described as models.[12] Given this, we'll consider the process of modelling in general to understand how a bias can become embedded in a model. A **model** "is nothing more than an abstract representation of some process."[13] Given certain parameters and a specific state of affairs, a model can allow us to predict what the outcome will be. For example, to decide what gift to give a loved one, you might create an informal model to help you predict what will be a good gift based on a careful balance of their interests, their needs, and your own capacities for obtaining gifts. We could create a model that attempts to predict future grade performance by considering factors like the time studied, whether someone went to a certain school, or even their financial background. As O'Neil explains, "All of us carry thousands of models in our heads. They tell us what to expect, and they guide our decisions."[14]

1.2.1 Modelling Assumptions

To better understand the issue of modelling and its ethical implications, imagine you have been asked to design an algorithm that can predict if someone has clinical depression. Machine learning requires lots of training data, so what metrics would you use as data to teach it to do this? According to the American Psychiatric Association, symptoms of depression might include severe feelings of sadness, a loss of interest or pleasure in activities once enjoyed, changes in appetite, trouble sleeping, loss of energy, feelings of worthlessness, difficulty thinking or concentrating, and so on. Despite the lengthy list of symptoms, there is no single unifying biological cause. For some patients there might be decreased activity in a neurotransmitter, while for others the symptoms might be caused by overactivity in the hormonal system.[15]

In other words, while we might use known symptoms of clinical depression to help determine if someone has clinical depression, it isn't something that we can directly measure on a biological level. If we want

---

12   Kearns and Roth 2020, 9.
13   O'Neil 2016, 18.
14   O'Neil 2016, 18.
15   Nemeroff 1988, 44–48.

to measure if someone has clinical depression, we must do so indirectly based on the empirically measurable symptoms that we think their depression is causing. If we could use these symptoms as metrics, we might be able to create a model that could predict if someone had clinical depression.

The process of defining the empirical measure of an abstract concept that is not directly observable is called **operationalization**. The electron, for example, is not something that we can directly measure or see, so how can it be empirically detected? The electron was discovered when positively and negatively charged plates were applied to a cathode ray. From the observable deflection of the rays, it was inferred that the particle had carried unobservable negative particles. Today, we can use wire chambers that detect subatomic particles and provide information on their trajectory by tracking the trails of gaseous ionisation. Such practices thus partially define how such unobservable particles can be operationalized and measured.

How should a sociologist operationalize a concept like human happiness? We can't directly measure everyone's happiness levels, so we might define empirical metrics that we can use. We could conduct a survey where we ask questions about happiness and compile those responses into a happiness index. But what if we can't reasonably expect to send out and get responses to surveys from that many people? In statistics, a researcher might try to substitute one measurement we would prefer (because they are direct or at least more direct) for a **proxy**. For example, to measure happiness we might create a statistical index based on well-known and publicly available data such as economic growth, poverty rates, or crime.

All of this raises an important question. What justifies us operationalizing a concept in a certain way or using certain proxies as stand-ins for the thing we are trying to measure? What justifies the connections we make between certain concepts and the phenomena that we take as evidence for those concepts for the purposes of making an inference? To answer this and gain insight into the ethical issues at stake, we can consider philosopher Helen Longino's account of evidential reasoning.

According to Longino, "states of affairs ... do not carry labels indicating that for which they are evidence or for which they can be taken as evidence."[16] If a child presents with red spots on their stomach we might

---

16　Longino 1990, 40.

take that as evidence that the child has chicken pox, but it is also possible that it is evidence of measles. The spots by themselves don't reveal which. According to Longino, "states of affairs are taken as evidence in light of regularities discovered, believed, or assumed to hold ... What explains why I come to believe [the child] has the measles rather than that, say, the moon is blue, is some belief that I have about the relationship between having a red-spotted stomach and having the measles."[17] These beliefs are what Longino calls background beliefs or **background assumptions**.

Background assumptions represent a regularity that allows us to take something as evidence for something else. If there was a large plume of smoke in the distance, you might infer the presence of a fire. Why? The background assumption "where there's smoke, there's fire" is likely operating as part of your inference. Longino points out that background assumptions can have all sorts of effects on the ways that we might take something as evidence of something else. Different background assumptions might enable us to reach the same conclusion, but for different reasons. One background assumption might have us pay attention to colour, while another might focus on location yet reach the same conclusion.[18] Alternatively, different background assumptions might lead us to reach different conclusions about the same state of affairs. Different aspects of the situation can be taken as evidence for competing conclusions. We thus need to consider what justification each side might have for their competing background assumptions.

INQUIRER'S TOOLBOX

2. Am I defining this problem too broadly or too narrowly?
7. How would I test any assumptions I have regarding the nature of the problem or a hypothetical solution?

Background assumptions connect a concept to the phenomena taken as evidence for it, and this means that the way we operationalize a concept or the kinds of proxies that we use as evidence for a concept we are measuring will be justified given whatever background assumptions we hold to be true or relevant. Sometimes these background assumptions can be directly confirmed.[19] For example, Kepler's laws of planetary motion hold that planets orbit in an ellipse. This generalization by Johannes Kepler

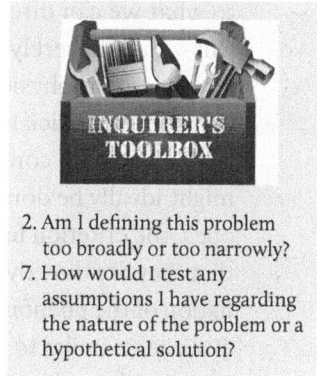

---

17 Longino 1990, 41.
18 Longino 1990, 42.
19 Longino 1990, 49.

was based on direct observations of planetary orbits made by Tycho Brahe in the sixteenth century.[20] However, as we have discussed, direct confirmation of a background belief isn't always possible.

If we can't directly confirm every background assumption that we might make use of, then we might believe them for bad reasons. As Longino notes, unless every background belief is directly confirmed then there seems to be no clear way to shield our reasoning about evidence from social and individual values and subjective preferences.[21] While we might be tempted to limit the scope of science and statistical reasoning to what we can directly confirm to bolster our certainty in the results, this would severely limit scientific investigation. It would mean, for example, that physicists would not be able to talk about unobservable particles. Statisticians would be limited in their ability to use proxies.[22]

We must be conscious of how science is done, rather than how it might ideally be done. Longino explains: "When a theory is being developed, the criterion for inclusion of specific hypotheses or principles is not that they are directly confirmed ... but that they are relevant to the explanation of the phenomena comprehended by the theory."[23] Notice that this attitude is similar to the account of inquiry presented in Chapter 1, which describes how thinking about information not only involves judgements about accuracy but also relevance to the problem at hand.

If we cannot directly empirically confirm every background assumption that we might use, then values and subjective preferences will play a role in determining what kinds of background assumptions you use. This will determine what kinds of evidence that you take as making a convincing case for something. Background assumptions may also govern how you understand the problem you are creating the algorithm for; they will affect how that evidence needs to be framed, how you will test your ideas and identify "success," and how you will deploy your results.[24] Thus, if an inquirer's background assumptions reflect biased attitudes, the models they create will reflect those biases as well. Therefore, considerations about background assumption are also ethical considerations.

---

20   Duhem 1954, 191.
21   Longino 1990, 48.
22   Longino 1990, 49–51.
23   Longino 1990, 51.
24   Biddle 2020, 4.

1.2.2 Values and Assumptions

Reconsider our attempt to operationalize human happiness. A statistician, unable to directly measure human happiness, might choose proxies such as economic growth, the amount of poverty, and the level of crime to create an index. What background assumptions are at play? One might be that "If an economy is growing, people will be happier" or that "If crime increases, people will be less happy." Notice two things about these assumptions. The first is that these assumptions might reflect capitalist values. For example, one assumes that people will be happier if they are wealthier or that economic growth translates into happiness. Secondly, these assumptions may not always be true; greater misery can sometimes follow economic growth. If we conceive of and measure happiness in this way, we may miss out on other aspects of happiness that we aren't measuring. This can lead to conclusions that reflect biases implicit in the background assumptions we have used.

We may sometimes want our values to influence our thinking. To understand this, reconsider making a model to detect clinical depression. We've discussed the fact that there is no single biological measure of depression, and that depression is measured according to a series of symptoms. One potentially relevant background assumption is whether we should understand it as a brain disorder. However, some researchers have argued that mental disorders should not be understood purely as brain disorders and that attempting to find correlations between symptoms and brain states to explain these symptoms is problematic. Instead, they argue that we should understand mental illnesses as a network of symptoms that interact with each other without assuming common biological causes.[25]

The central point is that if we assumed that we could operationalize clinical depression by picking a list of symptoms that correlate with brain states we think are associated with depression, this would only be justified given background assumptions about the relationship between mental health and biological events which are not directly verified. But what if we give up the assumption that there is a biological cause? This introduces yet another question. As philosopher Kristen Intemann asks, "if clinical depression is defined in terms of its symptoms, what criteria

---

25   Borsboom, Cramer, and Kalis 2019, 3–4.

should be used to distinguish clinical depression from the common blues?"[26]

Another way to understand Intemann's question is this: Without a single biological cause and given that patients suffer from a wide range of symptoms, how do we even know we are talking about a single illness? Intemann's answer is that what is common to all the symptoms of clinical depression is that they impair the functions most essential to human well-being such as eating, sleeping, and engaging in relationships. Depression thus impairs functions that are vital to a good life for humans. But what counts as a good life for humans is the value judgement that eating, sleeping, and engaging in relationships are central to a good human life. As she argues, "this value judgement (if justified) also justifies us in grouping together symptoms and cases of clinical depression. If we did not rely on this sort of value judgement, then cases of clinical depression would look naturalistically gerrymandered and we would be less justified in treating them as one disease."[27]

In other words, to operationalize a concept like clinical depression, we would do so given background assumptions that are themselves partially justified by value judgements. Such value judgements impact what we take as evidence for the concept, and thus whether we will identify the concept in the world and how. For example, to identify symptoms, a doctor might have to evaluate how much sleep is "excessive" or what counts as "abnormal" indecisiveness, or "appropriate" forms of guilt.[28] It should be noted that when values play this role in evidential thinking, the results will not always be ethically bad. In the case of clinical depression, for example, we might agree with Intemann's position that such a value judgement is justified because we do think that depression is harmful to human flourishing. Nevertheless, a background assumption could be invoked carelessly given values that reflect bias and prejudice.

Unfortunately, many harmful biases go undetected. The problem with background assumptions is that we aren't always conscious that we are using them. For example, if you check your speedometer in your car, you might take the dial as evidence that your car is moving at a certain speed. You may not consciously realize it, but such an inference only

26   Intemann 2001, S508.
27   Intemann 2001, S509.
28   Intemann 2001, S509.

makes sense if we assume various tenets of electromagnetism are true. The problem with commonly shared assumptions that sit in the background of our inferences is that they become invisible.

As Longino explains, when background assumptions "are shared by all members of a community, they acquire an invisibility that renders them unavailable for criticism. They do not become visible until individuals who do not share the community's assumptions can provide alternative explanations of the phenomena without those assumptions."[29] This is true in computer science; as Timnit Gebru explains, "The predominant thought that scientists are 'objective' clouds them from being self-critical and analyzing what predominant discriminatory view of the day they could be encoding, or what goal they are helping advance."[30]

1.2.3 Assumptions and Recidivism

This has been a long explanation of how values can become infused in a model and how it can lead to bias. A model requires that we understand some relationship between phenomena that we are using as evidence or data, and predicted outputs that we take to be examples of the thing we are measuring (Figure 8). Sometimes background assumptions are adopted for empirical reasons based on observed regularities, and sometimes they are based on practical reasons, such as an inability to directly measure something or a lack of resources permitting a better measurement. In machine learning, for example, operationalizing any concept means that you need large amounts of data that you can access for training purposes, and so you might be inclined to use proxies that are more public and accessible over other proxies that might be more relevant but less accessible.

As philosopher Justin Biddle explains, ML systems "are value laden in ways similar to human decision making because the development and design of ML systems requires human decisions that involve tradeoffs that reflect values."[31]

Let's return to the topic of predictive recidivism algorithms and why they generate bias. Recidivism is the tendency of a convicted criminal to reoffend. So, how would we operationalize the concept of recidivism

---

29 Longino 1990, 80.
30 Gebru 2020, 253.
31 Biddle 2022, 2.

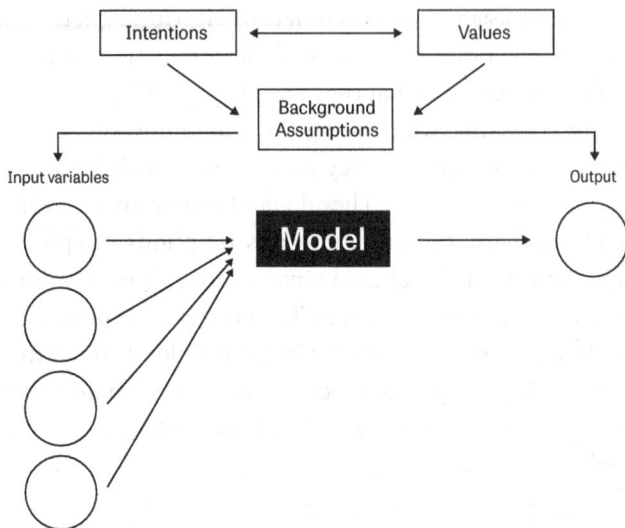

FIGURE 8 · The coherence of the model depends on background assumptions that justify the relationship between variables used as input data and what seems like an appropriate output from the model. The intentions of the developer and their values will influence what background assumptions they adopt. A background assumption may justify using certain variables as proxies for other properties or explain the meaning of an output value.

and how do we know if this will create a bias? You might think the most obvious predictor of recidivism is past criminal history as a safe background assumption. If someone has an extensive criminal history of repeat offences, they are more likely to offend again. But a single piece of information like this is hardly definitive. Many people have a past criminal history and don't reoffend, so if we rely on this intuition alone, we will label people as recidivists when we shouldn't. Some models, such as the "Level of Service Inventory-Revised" model, include lengthy questionnaires for the prisoner to fill out, determining attributes taken to be relevant to predicting if someone will reoffend.

If you were creating such a questionnaire, what questions would you ask? What things would you look for? Would you ask about the role that drugs and alcohol played in the crime? What background assumption are you relying on? As Cathy O'Neil points out, "Ask a criminal who grew up in comfortable suburbs about 'the first time you were ever involved with the police,' and he might not have a single incident report ... Young black males, by contrast, are likely to have been stopped by police dozens of times, even when they've done nothing wrong."[32] People of colour are far more likely, for example, to have police stop-and-frisk them than white people. If early 'involvement' with the police signals recidivism, "poor people and racial minorities look far riskier."[33]

The questionnaire also asks if friends have criminal records. If you are from a middle-class neighbourhood, the answer is likely going to be very different than if you are from a poorer neighbourhood. What this means is that if you think that having prior involvement with the police or having friends with criminal records (independent of context) is indicative of recidivism, then any algorithm based on those assumptions will naturally generate biased results. An algorithm fed with historical crime data "will pick out the patterns associated with crime. But those patterns are statistical correlations—nowhere near the same as causations. If an algorithm found, for example, that low income was correlated with high recidivism, it would leave you none the wiser about whether low income actually caused crime."[34]

Remember Longino's claim that states of affairs do not carry labels. The fact that you have friends who have a criminal record might be evidence that you are a recidivist, but it might also be evidence that you are poor. The fact that you have had prior contacts with police might mean that you will commit more crime, but it also might mean that police like to target people who look like you. If you are a prisoner, the algorithm is not evaluating whether you are likely going to be a recidivist, it is evaluating whether people it thinks are like you are likely going to be a recidivist according to whatever assumptions the creators had about recidivism. This makes it easier to understand how any stereotypes we have about recidivists can be transferred to an algorithm that will find those same stereotypical correlations.

The idea that a recidivism prediction algorithm must be "fair" also requires background assumptions that reflect certain values. What does it mean to be fair and how can we capture this concept in a statistically measurable way? At least two forms (there are many more) of "fairness" are represented by the concepts of "predictive parity" or "equalized odds." An algorithm will have predictive parity if it generates the same rate of true predictions as a fraction of all positive predictions regardless

32   O'Neil 2016, 25.
33   O'Neil 2016, 26.
34   Hao 2019.

of race. Equalized odds on the other hand holds that an algorithm is fair if it isn't more likely to generate false predictions for one group rather than another. Unfortunately, in some cases it is mathematically impossible to have an algorithm that satisfies both equalized odds and predictive parity.[35] Thus, it will be left to machine learning designers to judge which is the more appropriate form of "fairness."

Debates about competing definitions of fairness are important for how we evaluate real world use cases. For example, in 2016 an investigative report from ProPublica on the use of COMPAS revealed that the algorithm is biased against Black people. They found that Black defendants were 45% more likely to be assigned higher risk scores than white defendants after controlling for age, gender, and prior crimes.[36] In response, the corporation that owns COMPAS (now Equivant) argued that their algorithm is fair because it satisfies the criterion of predictive parity even though the ProPublica investigation used an equalized odds standard.[37]

Values and background assumptions can also affect how we understand the results. For example, the COMPAS algorithm generates probability scores of 1 to 10 where scores of 1–4 are considered "low," 5–7 are considered "medium," and 8–10 are "high." Declaring that what it means to have a "low" or a "high" score also reflects values about where we believe cut offs are for risk categories and how we think people with different scores should be treated. One of the background assumptions at work here is the idea that if someone has a higher score, they are likely to be a recidivist, and thus it is best that they remain behind bars longer.

If we reconsider what seemed like an obvious indicator of recidivism, a defendant's criminal history, in greater detail, we can understand how this can reflect biased values as well. If we use prior arrests to determine recidivism, then any biases that exist in pre-existing arresting procedures will bias the data as well. This is because "using arrests as a proxy for recommitting a crime means the algorithm is codifying biases in arrests, such as police officer bias to arrest more people of color or to patrol more heavily in poor neighborhoods."[38]

---

35 Chouldechova 2017.
36 Larson et al. 2016.
37 Biddle 2020, 14.
38 Ito 2019.

The fact that our values can, do, and sometimes must be used to operationalize a concept introduces important ethical implications. Who should get to decide what values are infused into the model through the use of background assumptions and how are they accountable for them? What sorts of ethical standards should a machine learning designer hold themselves to?

1.3 BIASED DATA

Developing physical colour photographs requires a delicate balance of chemical interactions in the layers of film that are sensitive to different colours of light. Getting accurate colour balance involves a careful balance of blues, greens, and reds. To create a standard for producing accurate skin tone colour, the print industry began using "Shirley cards" which often featured light-skinned Caucasian women in brightly coloured clothing as a basis for comparison (Figure 9). People with darker skin were not featured on these cards. The result is that colour photos could produce accurate skin tones for lighter-skinned people, but not for darker-skinned people.[39]

Shirley cards reflected European standards of beauty and often the most glaring problems were apparent in photos of multiple people with different skin tones.[40] The fact that white skin was treated as the "normal" skin tone allowed film technology to develop in ways that were biased against people with dark skin.[41] It wasn't until chocolate and furniture manufacturers complained that the film couldn't capture the proper colour of their products that improvements started to be made. A colour camera, like an algorithm, does not have personal or political preferences, but the way the technology

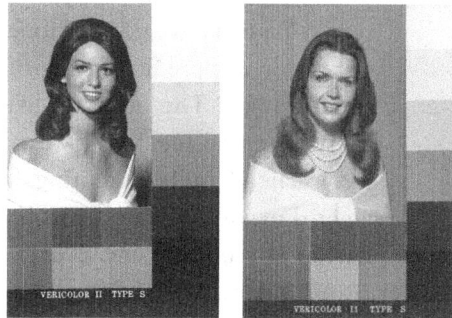

FIGURE 9 · SHIRLEY CARDS.
If you only used cards like these as a measure of colour accuracy in film, what is that likely going to mean when it comes to colour accuracy for other skin tones?

39   Lewis 2019.
40   Roth 2009, 115.
41   Ewart 2020.

INQUIRER'S TOOLBOX

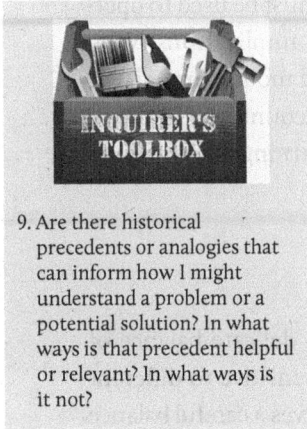

9. Are there historical precedents or analogies that can inform how I might understand a problem or a potential solution? In what ways is that precedent helpful or relevant? In what ways is it not?

was developed and the assumptions that governed what the "successful" use of the product was by humans created a technological bias.

The choice of background assumptions will affect how you understand the relationship between evidence (data) and the conclusion you want to reach (an output). They will determine what you choose as data, but it will also govern how you understand the quality of that data and what kinds of data you think are not relevant to a solution. If you are designing an algorithm for recidivism and you don't believe that white collar crime is a significant problem, then you won't mind if white collar crime is not featured prominently in your training data. Your interests in creating the algorithm will affect not only what you consider to be enough data, but whether that data is relevant to your goals.

An algorithm is going to be biased to the extent the training data used to produce the algorithm is itself biased. The test of any model created by machine learning is how well it works on new data that it wasn't previously trained with. If a recidivism algorithm is trained on prior arrests, and most arrests are from blue-collar crimes, then what would happen if we decided to reduce white-collar crime, and then used our recidivism prediction algorithm on these newly arrested white-collar criminals? The odds are likely good that the statistical profile of these potential recidivists won't match the profiles that the algorithm has learned based on training data from mostly blue-collar crime. The algorithm will generate errors because it was trained using a data set too narrow in scope to generalize to all crime.

A similar issue can occur in algorithms designed for predictive policing. The assumption is that you can more effectively deal with crime by sending police where they are most needed and making more efficient use of resources.[42] However, once again if the focus is on certain forms of crime rather than others, this can create biases. PredPol, for example, is a predictive policing algorithm that tries to predict crime in advance, but despite supposedly being blind to ethnicity, it has been accused of

---

42   Munn 2020.

being biased against poor, Black, and Latino neighbourhoods.[43] This is largely due to the algorithm only being trained to predict specific kinds of crimes rather than all crime. As we will see in the next section, this creates feedback loops.

Several prominent examples of biased algorithms stem from problematic training data. For instance, in 2018 Amazon shut down an automated hiring tool because it was found to be biased against women.[44] The algorithm had been trained on over ten years of résumés, most of which (not surprising in the tech sector) came from men. The result was that the algorithm penalized résumés that mentioned the word "women" (as in "women's chess club") and applicants from all-women's colleges.

Limited training data can also be potentially dangerous if an algorithm mislabels you. Facial analysis and recognition algorithms are being used in high-profile fields like security and policing. However, a study by Joy Buolamwini and Timnit Gebru on automated facial recognition reveals that many of the datasets used to train facial recognition algorithms contain significant demographic biases that tend to skew heavily towards white people and males, while darker-skinned people and women were underrepresented.[45] They found that the result is that darker-skinned females were likely to be misclassified with an error rate of 34.7% compared to an error rate of just 0.8% for lighter-skinned males.

The results of such errors in classification are serious. As Buolamwini and Gebru point out, "while face recognition software by itself should not be trained to determine the fate of an individual in the criminal justice system, it is very likely that such software is used to identify suspects."[46] People with darker skin are more likely to be stopped by law enforcement and be subjected to facial recognition searches. More importantly, Buolamwini and Gebru's study points to issues of **intersectionality** which involve overlapping systems of discrimination according to different social categorizations that make us overlook important sources of bias.

To understand this problem, consider the 1976 lawsuit by Emma DeGraffenreid against General Motors. She argued that GM discriminated against Black women but lost on the grounds that the company

---

43   Sankin et al. 2021.
44   Dastin 2018.
45   Buolamwini and Gebru 2018, 5.
46   Buolamwini and Gebru 2018, 1.

did not discriminate against Black people in hiring and so therefore did not discriminate against Black women. However, while GM did hire women, these were only for secretarial positions, and they did not hire Black women. Black men could be hired for a factory position, but a Black woman could not be hired as a secretary.[47] On the surface it can appear that there is no discrimination until we consider finer differences between different sub-groups.

Similar intersectional problems can occur within the data sets used for training purposes. For example, in assessing sources of bias within a database, Buolamwini and Gebru decided to use phenotypic labels (observable traits) because "phenotypic features can vary widely within a racial or ethnic category ... Thus, facial analysis benchmarks consisting of lighter-skinned Black individuals would not adequately represent darker-skinned ones."[48] By considering the range of different skin tones and gender classifications, they found that different biases emerge. Darker-skinned females were far more likely to be misclassified than darker-skinned men and even more so compared to lighter-skinned men. If our background assumptions aren't sensitive to intersectional issues, therefore, we open the door to bias.

Understanding the logistics of where your data is coming from and how it is collected is an important step in understanding sources of bias. A study by the Michigan Institute for Healthcare Policy & Innovation discovered that bias can creep into medical data banks based on recruitment efforts at the data collection stage. The Michigan Genomics Initiative was seeking blood from patients waiting for surgery and recruiters sought to approach all adults who had an intravenous line in place before their surgery to donate blood because it was more convenient to collect blood that way. The study found that this introduced a bias because the pool of patients was more likely to be older, white, and socioeconomically advantaged men compared to the general Michigan population.[49]

---

47   Crenshaw 1989, 144.
48   Buolamwini and Gebru 2018, 4.
49   Spector-Bagdaddy, Weins, and Creary 2021.

2. What Are the Ethical Consequences of Bias in Artificial Intelligence?

Failure to consider how background assumptions can translate into bias represents a failure to understand the background assumption as an end-in-view. It isn't always obvious that using a proxy for one concept can act as a proxy for another until we consider how things work in practice. Recall from Chapter 1 the example of an auto insurance algorithm. Because machine learning requires massive amounts of data, it is tempting to use certain variables as proxies simply because there is more public data available. Insurance companies, for example, will create an "e-score," or in other words their own proprietary rating system, by pulling in credit reports as well as demographic data. The e-score then stands as a proxy for a responsible driver.

Failure to pay your bills on time might suggest that you could be a risk on the road. Such a background assumption might justify using credit reports, but what about where you live? If you live in a neighbourhood where accidents are more common or where more people live who get into accidents, might this suggest that you could be a risky driver too? Let's say that we accept that this is good evidence to use, what biases might it introduce? The result of using demographic data and credit histories to produce e-scores as proxies for good drivers is that they also function as proxies for minorities and the poor. As Cathy O'Neil points out, "if the system attributes risk to geography, poor drivers lose out. They are more likely to drive in what insurers deem risky neighborhoods. Many also have long and irregular commutes, which translates into higher risk."[50] Thus, we must consider whether an assumption that might make sense introduces an unacceptable bias against the poor.

A similar problem occurs for PredPol, whose founder Jeffrey Brantingham stresses that the model is blind to race and ethnicity.[51] The model does not even consider the individual, only geography. The algorithm can be trained on violent crime, but a convenient source of additional training data can come from including non-violent crime (particularly if you also assume there is a relationship between these offences and violent crime). However, by focussing on crimes like

---

50   O'Neil 2016, 169.
51   O'Neil 2016, 86.

vagrancy, panhandling, and small quantities of drug sales, police are drawn to issues endemic to impoverished neighbourhoods. If we were to chart all such "nuisance crimes" visually, "such a crime map would track poverty."[52] In the United States, where cities are often segregated with certain districts correlating to different ethnicities, "geography is a highly effective proxy for race."[53]

The background assumption that your geography is relevant to you being a risky person and the assumption that people who live near you are like you are both required for such an algorithm to work, but once we understand how these proxies work in practice, we can understand how these proxies will import biases into an algorithm. Sometimes a failure to include enough data is simply due to a lack of better data resources, but often it is because of a failure to examine the justification of the background assumptions underlying our decisions about what kinds of information to include or exclude in machine learning. So, we should understand bias as a problem that not only occurs because of a lack of information, but as a problem that results from a lack of critical thought or as a deliberate effort to frame something in a biased way.

2.1 FEEDBACK LOOPS

A significant consequence of biased AI is the establishment of **feedback loops** where biased data reinforces the background assumptions that give rise to bias in the first place. To understand this phenomenon, let's reconsider the PredPol algorithm. A predictive policing model trained on data mostly featuring "nuisance" crimes instead of violent crimes will result in police being sent into predominantly Black and poor neighbourhoods.[54] Police sent to those neighbourhoods will not only report more crimes in those neighbourhoods by virtue of being there more often, but they will also make more arrests. This results in more data for the algorithm, which then predicts more crime, which in turn results in more police presence.

Based on the assumption that geography is a good predictor of crime, more police are sent into specific areas rather than others, which creates new geographic crime data reinforcing the idea that certain geographic

---

52   O'Neil 2016, 89.
53   O'Neil 2016, 87.
54   Gebru 2020, 257.

locations have higher crime than others and justifying a greater police presence in the area. As O'Neil argues, "The high number of arrests in those areas would do nothing but confirm the broadly shared thesis of society's middle and upper classes: that poor people are responsible for their own shortcomings and commit most of a city's crimes."[55] We can imagine, for example, the counter feedback loop that is also occurring in this situation. In areas where less crime is reported, the model will recommend fewer police be sent. This will result in fewer arrests made, and as new data is fed into the model, it will recommend sending fewer police to those areas.

Other cases we've discussed involve feedback loops. As Amazon's hiring algorithm was largely trained on the résumés of successful candidates that are mostly white and male, it will be less likely to select women or any marginalized communities. As Timnit Gebru explains, "The model selects for those in the nonmarginalized group, who then have a better chance of getting hired ... This generates more biased training data for the hiring tools, which further reinforces the bias creating a runaway feedback loop of increasing the existing marginalization."[56]

Recidivism models and/or risk assessments models using e-scores are also susceptible to feedback loops. This can again be a problem when background assumptions aren't questioned, and we begin to essentially understand the concept as it is defined by the statistical model. Our statistical understanding becomes our benchmark for understanding our world, but it is based on assumptions that don't reflect reality. Yet, the results of the model only serve to reinforce these same assumptions (Figure 10).

Consider a more visual thought experiment to demonstrate these points. In 2015 Google developed an algorithm called Deep Dream to recognize things in an image. Users could upload an image or video and the algorithm would produce a distorted hallucinatory effect where everything in the image would look like eyes or have dog faces. Why dog faces? The algorithm was trained on images from ImageNet, but instead of using the entire database, they only included dogs. The result is that "Google's Deep Dream sees dog faces everywhere because it was literally *trained* to see dog faces everywhere."[57] Now, imagine that we created

---

55   O'Neil 2016, 89.
56   Gebru 2020, 257.
57   Brownlee 2015.

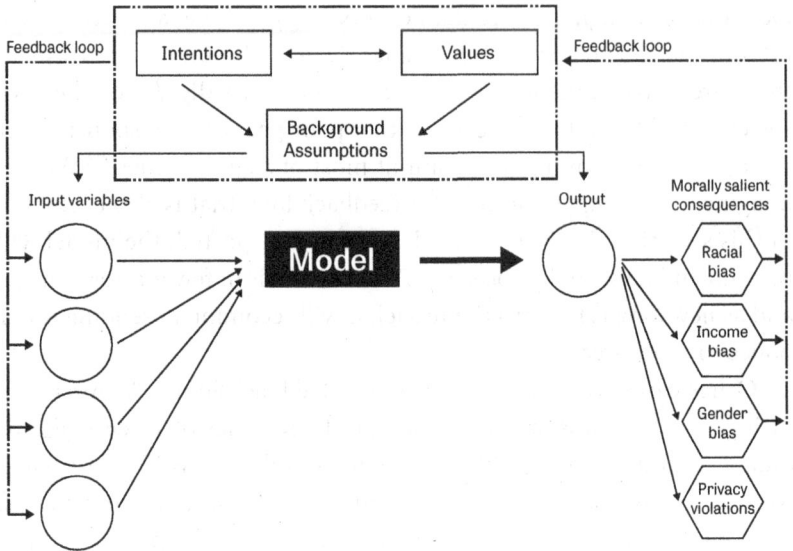

FIGURE 10 · The output of an algorithm will translate into real world consequences. Some consequences can be morally problematic, such as when an algorithm contributes to gender bias or enables systematic bias against racial minorities. Another example of a morally salient consequence could be a violation of individual privacy (see Chapter 6). The effects of a biased algorithm will create new data for that algorithm that only serve to reinforce the same biased assumptions, resulting in a feedback loop.

an image recognition algorithm based on the assumption that the only things that exist are dogs. We then only use dog training data, and the algorithm only sees dogs.

The problem is that when people begin to treat the algorithm as a tool to tell us what the world is like, it will reinforce the idea that only dogs exist. An algorithm is only as good as the assumptions that underlie it and the data used to train it. It's a system of garbage in, garbage out. Even when our assumptions are more solid, obtaining the data required or operationalizing a concept for specific contexts can mean making difficult and complicated trade-offs, and if we forget this, we run the risk of overgeneralizing the utility and accuracy of an algorithm and mistaking statistical representations of reality for reality itself. We will consider this theme in more depth in Chapter 6.

2.2 ETHICAL RESPONSIBILITY?

Given what we've discussed, there are many factors that can contribute to a biased algorithm and there are many consequences that can follow from their use. If you are a developer you have to deal with background assumptions and values that you may not be fully conscious of; you may have to sacrifice accuracy by settling for questionable proxies simply because it's the only way to get the large amount of training data required, and your decisions might have spill-over effects in the form of a feedback loop that can not only reinforce bad thinking on your part but help marginalize and otherwise harm whole groups of people, even if that wasn't your intention. Given all of this, it is reasonable to ask: To what extent is an AI developer ethically responsible for these things?

If we reflect on the previous chapter, we can attempt an answer. Recall from Heather Douglas that an AI developer is ethically responsible not to act negligently or recklessly. While we might ethically condemn the corporation that develops a biased algorithm that is designed to prey on its customers, what about the developer who creates long-term feedback loops involving an interplay of various sociological factors that makes life for marginalized people worse? As Douglas argued, scientists have no special foresight: "We cannot expect scientists to be seers."[58] This point is echoed by Gebru et al., who note that "Dataset creators cannot anticipate every possible use of a dataset, and identifying unwanted social biases often requires additional labels."[59]

Nevertheless, the ethical consequences of bias are not unknown, and given that we understand feedback loops and how they work, it stands to reason that if someone is developing an algorithm, they shouldn't do so in a way that would recklessly or negligently harm the public. If someone operationalizes a concept using empirically questionable or outdated background assumptions without considering how such a conception would play out in the real world or who it might negatively affect, that is negligence. If an AI developer substitutes a reliable proxy for a metric used for an for e-score simply because there is more data available to complete the task and despite creating a feedback loop in their own algorithm, that is recklessness.

---

58   Douglas 2010, 67.
59   Gebru et al. 2021, 92.

To avoid this, an AI developer, like the moral inquirer from Chapter 1, will need to consider actions as ends-in-view. Considering your actions as an end-in-view means considering your goal in terms of the means required to obtain it and the consequences that this will likely lead to. We may not expect an AI developer to be a seer, but we would expect them to consider the likelihood of creating a feedback loop once they are aware of the possibility, and we would expect them not to charge ahead recklessly if they are aware of the risks. Understanding that there are moral consequences to your actions that you may not know about only reveals your responsibility to better understand the consequences of your actions before acting.

Recall the AI developer who creates an algorithm to assess risk for the purposes of determining insurance rates. They choose to use credit scores, geographic history, and various other factors and produce an e-score to use as training data for the algorithm. As a result, poorer people pay disproportionately more for their insurance even though their driving history only makes up a tiny portion of their estimated risk. If we rely on geography to determine rates and there are correlations between geography and driving risk, the model will increasingly rely on geography rather than on personal driving history to determine risk, creating a feedback loop and making it easier for riskier drivers living in less risky (wealthier) neighbourhoods to pay less simply because of geography.

We can understand that the developer's goal is to be able to determine appropriate insurance premiums quickly and efficiently. To achieve this, the developer builds a model using machine learning and feeds in data to train that model. These are the means used to achieve the developers' ends and reflect the available resources. Although this gives us a slightly more nuanced understanding of their goal, remember that an end-in-view is not only understood in terms of the means required to obtain it, but the consequences that will likely occur. Once we work out our goal as an end-in-view, we understand that we are not merely calculating risk, but are doing it in a specific way and producing specific consequences. The original goal takes on a more complex meaning because we understand the effects it produces.

Assume that our developer isn't just in it for the money and that as a general ethical goal they do not wish to exacerbate income inequality. By working out their goal to calculate risk in terms of its consequences

End-in-view

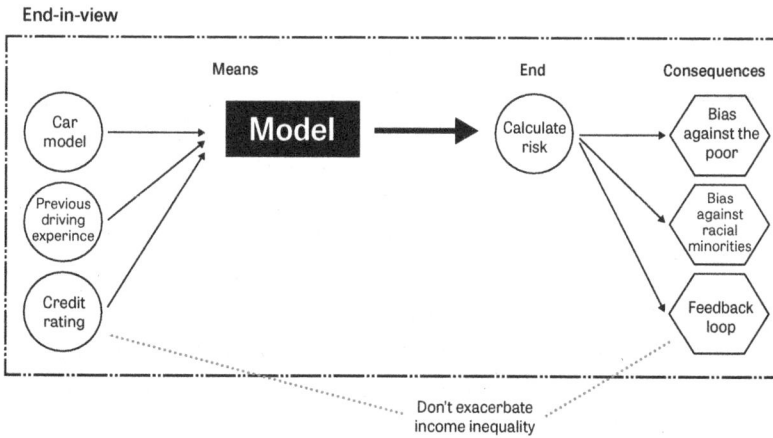

FIGURE 11 · Recall from Chapter 1 that understanding our actions as ends-in-view can yield ethical insights. When we consider the consequences of proceeding in just this way and understand our end as an end-in-view, we appreciate that the means used (in this case credit reports) contributes to a problem that is at odds with our other end goal of not exacerbating income inequality.

and not just as an end in itself, they come to realize it conflicts with their goal of not making income inequality worse (Figure 11). As noted in Chapter 1, it is this conflict of ends that will give rise to moral reflection.

However, with greater understanding the developer can anticipate these consequences as an end-in-view and can now reconsider alternative approaches. As Dewey explains,

> When things are weighed as means toward that end, if it is found that it will take too much time or too great an expenditure of energy to achieve it, or that, if it were attained it would bring with it certain accompanying inconveniences and the promise of future troubles, it is then appraised and rejected as a 'bad' end.[60]

Formulating ends-in-view allows us to better deliberate about our options. Moral deliberativeness "when it is habitual is conscientiousness.

---

60   Dewey 1939, 24.

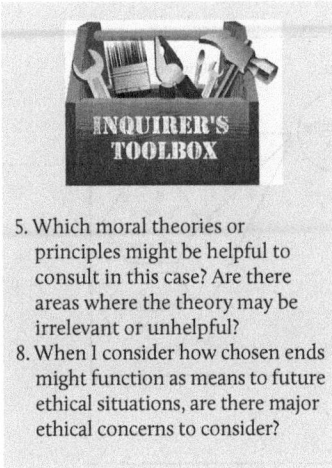

INQUIRER'S TOOLBOX

5. Which moral theories or principles might be helpful to consult in this case? Are there areas where the theory may be irrelevant or unhelpful?

8. When I consider how chosen ends might function as means to future ethical situations, are there major ethical concerns to consider?

This quality is constituted by scrupulous attentiveness to the potentialities of any act or proposed aim."[61]

In other words, an AI developer is ethically responsible to the extent that they have developed a level of conscientiousness about their own work such that they can clearly formulate ends-in-view that are not morally reckless or negligent. Let us now look at some additional case studies and close by considering potential solutions to help the AI developer be more conscientious and reduce the harmful effects of biased AI.

3. Case Studies of Bias

Before investigating how to mitigate bias in machine learning, let's consider the real-world effects that biased AI has on the world. Using our tools of inquiry, we will see if we can better understand these issues given what we have learned.

3.1 TARGETED ADVERTISING

Algorithms are incredibly useful for advertising. Imagine being able to correctly predict the kind of person you are, the wants you have, and then advertise exactly the kinds of things that you (or people like you) are likely to purchase. An algorithm can identify different kinds of shoppers and show them relevant advertisements. Unfortunately, advertising in this way can also perpetuate biases magnified by the sheer scale available for automated advertising. Sometimes this is due to what is being advertised and sometimes it's based on who is being targeted by the advertisement.

Many online advertisements use automated gender recognition to detect the gender of a potential shopper and use that to help target

---

61   Dewey and Tufts 1932, 299.

advertisements.[62] This can be problematic for several reasons. For instance, such a system can perpetuate gender stereotypes. The algorithm would have to be trained using background assumptions regarding gender to allow it to differentiate one gender from another, which would govern what kinds of advertisements should be shown to a consumer based on their gender identity. Unfortunately, these assumptions often simply reflect stereotypes. Urban Outfitters attempted to personalize their website for shoppers according to the perceived gender of the user, with certain items only being recommended based on gender.[63] Schemes like this can perpetuate the notion that certain things are only for women or only for men (Figure 12).

Automated gender recognition can be problematic if they treat gender as a static concept. Gender presentations vary across different cultures and often these algorithms are trained with data that includes few or no transgender or nonbinary people. As Gebru notes, the effects can be severe, ranging from the algorithm misgendering someone to potentially outing them in public.[64] Such cases demonstrate that bias stems from problematic background assumptions about gender and biased data that doesn't reflect the diversity of gender identities that exist in the world.

As previously discussed, bias can also occur if there is a deliberate effort to discriminate. For example, companies like Facebook or Google can help sell advertising to specific consumers based on who they are. In 2019, the United States sued Facebook for

FIGURE 12 · Bias in advertising is not new. The above Hoover advertisement clearly perpetuates the idea that housework is a woman's chore and hence women would be happier with the additional convenience of a vacuum cleaner. The difference now is that AI can target you based on your gender and send you similar kinds of advertisements that also perpetuate specific gender stereotypes.

62   Vincent 2021.
63   Singer 2012.
64   Gebru 2020, 259.

violation of the Fair Housing Act for engaging in housing discrimination by allowing advertisers to restrict who can see certain ads based on characteristics like race.[65]

3.2 FACIAL RECOGNITION

Facial recognition algorithms present several ethical problems. We will discuss privacy concerns in Chapter 6, but there are also problems with bias. As usual, we need to consider what these algorithms are being used for, what assumptions go into them, and the quality of training data. For example, in Chapter 2 we discussed a case where a facial recognition algorithm was likely being designed for the purpose of targeting certain segments of the population. The use of facial recognition for criminal justice purposes can be subject to abuse if governments start targeting citizens.

In 2020, IBM, Microsoft, and Amazon announced that they would end sales of their facial recognition technology to police due to concerns that it would be used to promote racial discrimination.[66] As noted, Black and brown people are disproportionately targeted by police and there is a long history of disproportionate police surveillance of Black and brown neighbourhoods. This means that these groups are more likely to have facial recognition technology used against them and that their mugshots will be used as training data for such algorithms.

Nevertheless, Buolamwini and Gebru have shown that facial recognition algorithms are far more likely to misidentify darker-skinned people over people with lighter skin and to misidentify women over men.[67] Black women were misidentified almost 35% of the time. A federal study also suggests that these algorithms are more likely to misidentify Asians, Native Americans, and Pacific Islanders as well.[68] The use of facial recognition as evidence is thus problematic, and the higher rate of false positive identifications for Black and brown people means that they are more likely to be targeted for crimes they did not commit.

Facial recognition is also used by employers. In 2021, Pa Edrissa Manjang sued Uber for its use of facial recognition software, which

---

65   Benner, Thrush, and Issac 2019.
66   Crockford 2020.
67   Buolamwini and Gebru 2018, 9–11.
68   Harwell 2019.

drivers considered to be racist.[69] Uber introduced the automated system to check their drivers' ID. However, Manjang lost his job when his account was deactivated after the software failed to recognize his face several times. Microsoft designed the software, but as noted it was very unreliable for darker-skinned people, and its inaccuracies were owing to imbalances in the training dataset.[70] As we discussed earlier with Gebru and Buolamwini, a lack of background assumptions regarding the importance of variations in skin tones likely helped justify the use of too small a training dataset relative to the population at large.

3.3 LANGUAGE PROCESSING AND TRANSLATION

Machine learning can be applied to language processing, language recognition, and language translation. For example, an algorithm might be trained to suggest synonyms or additional words for a partial sentence. However, these applications can also be biased. One way in which an algorithm can learn the meaning of a word is to be trained on how that word is embedded in language; how words with similar meanings tend to occur in specific places in our language, and how differences in these occurrences can point to relationships between meanings. Using this principle, an algorithm might learn word relationships and predict that the word "queen" follows from a claim such as "man is to king as woman is to ..."

Unfortunately, training using word embedding requires massive amounts of data that can introduce biased results. A study by Bolukbasi et al. using the word-embedding algorithm word2vec and trained on Google news articles with over three million words was found to exhibit gender bias by reproducing gender stereotypes.[71] For example, it would complete the analogy "man is to computer scientist as woman is to ..." with "homemaker." Women were considered more likely to be librarians, nannies, nurses, or receptionists, while men were more likely to be considered financiers, philosophers, captains, or the boss.

If we look at language translation, there are similar problems. There are countless examples from Google Translate that feature translations that reflect gender stereotypes.[72] If you translate the sentence "She's a

---

69   Butler 2021.
70   Castelvecchi 2020.
71   Bolukbasi et al. 2016, 3.
72   Kayser-Bril 2020.

computer scientist" from English into Malay, for example, then translate that sentence back into English, you will get "He's a computer scientist." Poor translation can have horrific consequences. A Palestinian was arrested by Israeli authorities for writing "good morning" in Arabic after Facebook Translate translated it into Hebrew as "attack them." As Gebru points out, these translation algorithms are far more accurate when it comes to translating English into French. The lack of input from Palestinian people and from other Arabic-speaking populations means that mistakes in translation are more likely to occur.[73]

3.4 MEDICAL APPLICATIONS

Algorithms can be used in a wide range of applications in the medical field, everything from helping to diagnose a patient to determining what kinds of care a patient should receive. An algorithm that predicts healthcare risks is used to help make managing patients more efficient and effective by recognizing patients that would benefit from "high-risk care management" programs that provide chronically ill individuals with access to special staff and better care. By identifying patients that require these services sooner, the hope is to reduce the risk of serious complications that might be more costly in the long run.

However, the algorithm needed a proxy to measure risk and one assumption was that a patient's prior healthcare spending would indicate higher risk and greater need for care. Cost is also a widely available piece of data to work with. However, a study of the algorithm revealed that it was subject to racial bias.[74] The care provided to Black people costs less than for a white person with the same number of chronic health problems. Researchers suggest that this may be because people of colour have lower incomes, and even when they have insurance tend to use medical services less frequently. Racial biases in the medical system can also contribute to a lack of trust and lead to lower standards of care.[75]

In other words, for the same cost a Black person is substantially sicker than a white person.[76] Because the algorithm predicted need based on cost, and because of unequal access to healthcare, Black people were

---

73   Gebru 2020, 265–66.
74   Obermeyer et al. 2019.
75   Vartan 2019.
76   Ledford 2019.

assigned lower risk scores than white people and were thus given inadequate care. Failure to carefully consider the background assumptions and the consequences they could produce worsened existing biases in the data. We must be aware of existing biases in terms of the proxies used to understand that data.

4. Suggested Solutions?

It is obvious that biased AI can cause ethical problems. However, we've also discussed how complicated it can be to avoid bias. Machine learning requires massive data and sometimes that means that the only data available in such quantities may involve implicit or explicit biases. Even if a developer doesn't consciously want their algorithm to be biased by including factors like race or gender, often certain variables can act as proxies for these factors, thus echoing biases that already exist in the world. So, what kinds of solutions can mitigate bias? Should we focus on technical solutions for the problem, or should we look elsewhere?

4.1 TECHNICAL SOLUTIONS

Earlier we discussed different technical definitions of "fairness" that a developer might use as a standard to determine if their algorithm is biased. We've discussed the concept of fairness as *equalized odds* and *predictive parity*, but a 2018 survey identified twenty different statistical definitions of fairness that have been articulated by scholars that measure slightly different things.[77] Some of these focus only on the predicted outcomes, some on the predicted outcome compared to the actual outcome, and others compare actual outcomes to predicted probability scores.

For example, one definition known as *well-calibration* defines fairness between protected groups and unprotected groups as an equal probability that either group belongs to a predicted class. We may wish to ensure, for instance, that on recidivism scores your race or ethnicity alone does not make you more likely to be considered a recidivist. On the other hand, we may want to adopt a standard called *balance for the*

---

77   Verma and Rubin 2018, 2.

positive class, where the average score received by people constituting positive cases should be the same for protected and unprotected groups. Or we may want to *balance for the negative class*, where we ensure that the average score received by people constituting negative cases should be the same for protected and unprotected groups.

Different standards may yield different results and thus may be more or less appropriate based on what moral consequences the developer is most concerned about. Some may be more concerned about preventing **false positives** while others may be more concerned about **false negatives**. More importantly, these standards can't always be satisfied at once. A 2016 paper demonstrated, for example, that except in very specific circumstances, it is not mathematically possible for a model to be well-calibrated, balanced for positive cases and balanced for negative cases simultaneously.[78] In other words, even when we try to build in technical definitions of fairness, we still need to be conscious of what background assumptions inform these decisions and what ethical trade-offs they carry.

Some corporations responsible for developing these algorithms have also begun to develop toolkits for mitigating bias. For example, IBM has developed Al Fairness 360, which features several metrics to allow for the examination, reporting, and mitigation of bias.[79] IBM has also developed tools to improve both transparency and bias such as IBM Watson OpenScale, which provides real-time bias detection and aims to improve explainability to make Al predictions more transparent.[80] While solutions like these may help with biased Al, they won't address the deeper social issues that can lead to biased data and biased modelling in the first place.

4.2 SOCIAL SOLUTIONS

Some major causes of biased Al include using background assumptions or datasets without fully understanding how they work in practice. Thus, an important step in finding solutions for the problem of biased Al is going to include increased transparency and accountability. As mentioned by Gebru, the notion that scientists are "objective" can be an obstacle to self-criticism. Science is attracted to the notion that it seeks

---

78   Kleinberg, Mullainathan, and Raghavan 2016, 6.
79   IBM Developer Staff 2018.
80   IBM 2022.

to discover truths of the world that transcend our human perspectives, to discover what the world is "really" like. This is sometimes called an Archimedean point since it is the perspective that someone would have if they could stand outside of the world and observe it independently. For similar reasons, it is also called a God's-eye view of the world.

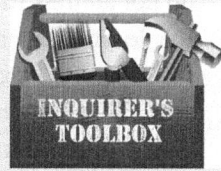

INQUIRER'S TOOLBOX

12. Are there multiple solutions to this ethical problem?

Philosopher Thomas Nagel called this detached objective perspective "the view from nowhere."[81] It is obvious why such a view is attractive: it seeks to transcend our own perspectives that might get in the way of discovering something true. On the other hand, if our beliefs do reflect our perspectives or if there is no way to represent something without adopting some perspective, but we think we are adopting a view from nowhere, it can prevent us from criticizing our own thinking.

Mathematics enjoys prestige because, the thinking goes, so long as we are just calculating we are being objective. However, as discussed, background assumptions are embedded in statistical modelling that reflect different perspectives. Gebru argues that education in science will have to move away from the view from nowhere. She notes that "the hardest thing to change is the cultural attitude of scientists. Scientists are some of the most dangerous people in the world because we have this illusion of meritocracy and there is this illusion of searching for objective truth."[82]

As many who have studied the problem of bias in AI will affirm, there needs to be greater inclusion in development. As Gebru notes, "Ethical AI is not an abstract concept but is one that is in dire need of a holistic approach. It starts from who is at the table, who is creating the technology, and who is framing the goals and values of AI."[83] With greater diversity of critical feedback, problematic background assumptions are more likely to be detected. Dominance of certain groups "and underrepresentation of others in natural language processing, computer vision, and machine learning ensures that the problems these groups work on

---

81   Nagel 1989.
82   Smith 2019.
83   Gebru 2020, 264.

do not address the biggest challenges faced by those who are not part of the dominance group in that field."[84]

Greater awareness and critical review of background assumptions requires greater transparency about what background assumptions are being used. Recall Longino's point that background assumptions that are shared by everyone become invisible to inquirers and immune from criticism. The view from nowhere encourages the same thing because beliefs that actually do reflect specific perspectives are represented as if they aren't there, rendering them invisible. As O'Neil explains, "many poisonous assumptions are camouflaged by math and go largely untested and unquestioned."[85] We don't notice these invisible or camouflaged assumptions until we are presented with alternatives.

Does focussing on background assumptions, which may reflect our social and ethical values, mean giving up on objectivity? Longino doesn't think so. She argues that "objectivity has to do with mode of inquiry ... [it] is to claim that the view provided ... is one achieved by reliance on nonarbitary and nonsubjective criteria for developing, accepting, and rejecting ... hypotheses and theories."[86] We can determine nonarbitary and nonsubjective criteria for theory acceptance by adopting a normative framework that encourages transformative constructive criticism from a diverse range of perspectives.

According to Longino, "To say that a theory or hypothesis was accepted on the basis of objective methods does not entitle us to say it is true but rather that it reflects the critically achieved consensus of the scientific community."[87] As she explains, "A method of inquiry is objective to the degree it permits transformative criticism."[88] She articulates four criteria necessary for any scientific field or community to achieve transformative criticism:

Recognized Avenues for Criticism. Peer review is a standard avenue for scientific communities to engage in transformative criticism to recognize and weed out idiosyncratic or otherwise faulty assumptions. This means that there must be public forums such as journals or conferences

---

84   Gebru 2020, 266.
85   O'Neil 2016, 7.
86   Longino 1990, 62.
87   Longino 1990, 79.
88   Longino 1990, 76.

to permit such peer review. Importantly, it also means that there should be equal priority for critical review as there is for original research.

Shared Standards. Criticism must be relevant and understandable if it is to be useful. Thus, to facilitate criticism, a scientific community must be bound to public standards governing issues like empirical adequacy, truth, and other substantive principles of the field. As we noted in Chapter 1, however, we may have a while to go before shared standards can be understood across the entire AI field. Alternative perspectives and criticism should have some bearing on the concerns of the scientific community, but the community is also responsible for changing standards in response to criticism.[89]

Community Response/Uptake. For a community to make improvements, there must be uptake of criticism received. Scientific and engineering communities must change in response to critical discussion, and this should be reflected in textbooks and other publications, as well as in grant allocations. However, this doesn't mean that anyone who is facing criticism must recant; the question is whether they can defend their work and their assumptions in response to criticism.

Equality of Intellectual Authority. A community is unable to be objective if certain sets of assumptions dominate by virtue of the political and social power of its adherents. Ideas should dominate because they are empirically successful and can survive rigorous criticism, not because alternative ideas are excluded from consideration. As Longino notes, "The exclusion, whether overt or more subtle, of women and members of racial minorities from scientific education and research has also constituted a violation of this criterion."[90] Instead, as many viewpoints as possible must be offered a seat at the table and have their perspectives and criticisms heard.

Objectivity emerges because individuals can participate in a collective give-and-take of critical discussion. As Longino explains, "objectivity is dependent on the depth and scope of the transformative

---

89   Longino 2001, 130.
90   Longino 1990, 78.

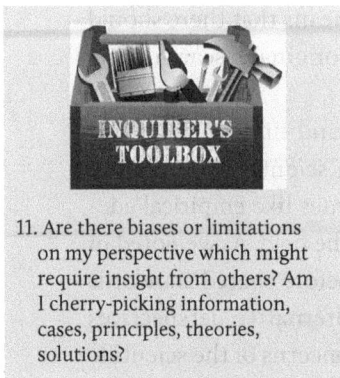

INQUIRER'S TOOLBOX

11. Are there biases or limitations on my perspective which might require insight from others? Am I cherry-picking information, cases, principles, theories, solutions?

interrogation that occurs in any given scientific community. This process can ensure that the hypotheses accepted as supported by some set of data do not reflect an individuals' idiosyncratic assumptions about the natural world."[91] Critical consensus cannot be reached by arbitrarily excluding the perspectives of other members of the community. We must ensure that our assumptions are not being shielded from critical scrutiny.

The intent is to expose ideas to the broadest possible range of criticism. But this also creates responsibilities for the community. A community "must also take active steps to ensure that alternative points of view are developed enough to be a source of criticism and new perspectives."[92] Not all communities fully meet these standards, but only to a degree. Nevertheless, if AI development communities were to follow these standards, this would help ensure greater transparency about the background assumptions we use, as they would no longer be immune to criticism because they are shared by everyone.

Increasing transparency in AI development is also a potential tool in finding a solution to biased AI. In a paper entitled "Datasheets for Datasets," Gebru et al. propose using datasheets, like the ones used in electronics, on datasets being used for machine learning, to increase transparency over how those datasets were created and what they are designed for (Figure 13). They note that "there is currently no standardized process for documenting machine learning datasets. To address this gap, we propose datasheets for datasets."[93]

A datasheet is a document used in electronics that describes the operating characteristics, performance results, recommended usage, and other information, so that users and engineers understand what the product is and the role of components in the overall system. A similar datasheet could be produced for every dataset which would document its motivation, collection process, recommended uses, and so on. The

---

91   Longino 1990, 79.
92   Longino 2001, 132.
93   Gebru et al. 2020, 86.

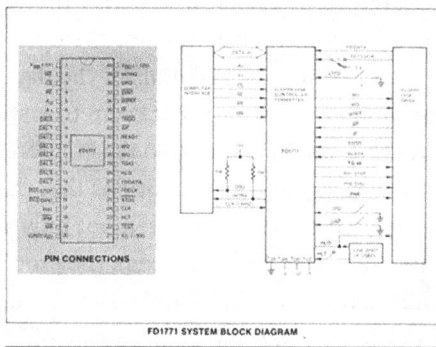

FIGURE 13 · A datasheet like this one has been used in consumer electronics for decades. It describes the operating characteristics and expected results and contains recommendations for usage. Is it possible that creating datasheets for datasets could introduce greater transparency and accountability?

hope is to increase transparency and accountability within the machine learning community to help mitigate biases and improve reproducibility.

"Datasheets for Datasets" outlines a series of questions intended to draw out the information that a datasheet for a dataset might contain. Some questions focus on the motivations for the creation of the dataset, such as what its intended purpose was, who created it, and who funded its creation. Some questions are about data collection, such as what mechanisms were used to collect the data, or whether an ethics review was conducted. Other questions ask a potential dataset creator whether they consider it a complete set or a sample or whether there are sub-populations, while others ask questions regarding what the dataset was envisioned for or whether there is anything about the composition of the dataset that might impact future use.

If technical documentation were required in this way for all the components of machine learning, it would help introduce transparency and assist potential users to realize their long-term consequences. As Gebru et al. note, "For dataset creators, the primary objective is to encourage careful reflection on the process for creating, distributing, and maintaining a dataset."[94] Is it possible that requiring transparency in the form of documentation, combined with the standards of transformative criticism, could help prevent the worst excesses of biased AI?

Imagine, for example, if corporations and the governments who purchase them had to publish datasheets for their recidivism algorithm that detailed not only the technical details of the datasets, but also the values and background assumptions that seem to justify using datasets in the specific ways they are being used to reach inferences, particularly in light of alternative background assumptions that they may have been aware of. Imagine if such documents noted not only who was involved in development, but also who was involved in the process of transformative criticism required for refinement and what alternative background assumptions the developers may have been aware of.

In other words, finding solutions to the problem of bias in artificial intelligence is going to require that we take steps to be more transparent about our thought processes, acknowledging the value-laden and contextualized nature of our background assumptions, and subjecting those assumptions to the broadest range of criticism possible to ensure that whatever is developed does not reflect idiosyncratic beliefs or overlook long-term ethical consequences that may not be otherwise apparent to the developer. Only then can a developer be conscientious about their choices and form clear ends-in-view, as opposed to acting recklessly or negligently.

5. Conclusion

Much could be said about artificial intelligence and bias. However, we now have a basic understanding of how machine learning works and how values can become embedded in algorithms developed through machine learning. Sometimes AI is biased because it is designed to be,

---

94   Gebru et al. 2021, 87.

sometimes because of problematic background assumptions that govern how we understand the relationship between data and our conclusions, and sometimes we have problematic data that is either biased or reflects bias because variables can act as proxies for other variables.

Resolving bias will require analyzing which perspectives are being reflected in our thinking and subjecting those ideas to broad criticism. It will also require us to make practical trade-offs. For example, many algorithms require so much data that finding non-biased sources may be impossible. So, do you try to mitigate that bias as best as you can based on your best guess of what a fair result should be, or do you give up on the algorithm altogether for being too morally dangerous?

This raises important questions. Instead of trying to fix the problem, should we simply ban the use of AI in specific cases such as in the criminal justice system? Are technical solutions sufficient? How can we make our background assumptions more transparent? What definitions of fairness are the most appropriate? How should the AI field be structured to permit greater transformative criticism? Should our response to the problems of bias in AI focus on changing AI or on changing society? To what extent should AI results reflect the world we wish to live in versus the world that we do live in?

ADDITIONAL MATERIAL

Equality • every individual or group is treated the same and given the same opportunities.

Equity • individuals or groups are given what they need to succeed.

Bias • a tendency for something to go in one direction rather than another, to favour certain conclusions rather than others, or to seek certain outcomes rather than others.

Machine learning • a type of artificial intelligence in which the algorithm learns from and improves through experience.

Supervised learning • a type of machine learning model that is built on labelled training data with known outcomes; the model is used to predict outcomes for unlabelled data with unknown outcomes.

Unsupervised learning • a type of machine learning model that finds hidden patterns within a set of unlabelled data; the model groups data that have similar properties.

Regression • a statistical technique that determines the relationship between variations of dependent variables to one or more independent variables. It determines the strength and character of the relationship between changes in independent variables and the quantitative effect on dependent variables.

Independent variable • variables that we believe will influence a dependent variable.

Dependent variable • variables that are influenced by independent variables.

Model • an abstract representation of real-world process(es).

Operationalization • the process of converting abstract concepts into observable and measurable phenomena.

Proxy • a variable that stands in place of an unobservable or immeasurable variable.

Background assumption • an assumed regularity or generalization that permits one to use a given phenomenon as evidence for other phenomena.

Intersectionality • examining overlapping systems of discrimination by race, gender, sexuality, disability, ethnicity, and other socio-political classifications as they relate to discrimination.

Feedback loop • when biased data reinforces the background assumptions that created the initial bias.

False Positive • an instance where a test incorrectly suggests that something is the case when it isn't. False positives are also known as a type I error in statistics.

False Negative • an instance where a test incorrectly suggests that something is not the case when it is. False negatives are also known as a type II error in statistics.

---

1. What is the significance of the "Shirley Card" as it pertains to the creation of biased AI?

2. We've discussed several examples of statistical definitions of fairness, including predictive parity, equalized odds, well-calibrated, balance for the positive class, and balance for the negative class. What reasons might prompt us to adopt one standard rather than another? Do these statistical definitions of fairness sound fair?

3. Given that AI can exhibit the same biases as humans, are there certain applications that should ban the use of AI? Is using human judges only a better alternative?

4. How can the field of artificial intelligence come to adopt the necessary shared standards to permit transformative criticism?

5. Will greater transparency and criticism from a diverse range of perspectives be sufficient to prevent the worst excesses of AI or will financial incentives outweigh any criticisms raised?

6. Read the article by scanning the QR code above. In groups, try to identify the various background assumptions that are being invoked in relation to the Anthropocene epoch. Do they make sense?

7. Are there some uses of AI where we should prefer AI to reflect pre-existing social biases?

Opacity, Risk, and Error

In November 2022, the AI research laboratory OpenAI released their AI-powered chatbot ChatGPT. ChatGPT is capable of mimicking human conversation, even the styles of specific individuals, and can be used for everything from composing music to writing a paper. Many critics are beginning to openly wonder if this is the death of the college essay, since a student could simply use ChatGPT to write their paper.[1] However, algorithms have also been developed to detect content written by ChatGPT. Imagine that you are a university professor and an algorithm tells you that a student of yours plagiarized by submitting an assignment written by ChatGPT. The student swears that they wrote their paper and that the algorithm must be wrong. What do you believe?

What if you don't even know what the algorithm was looking for or why it thinks the student cheated? How can you justify holding the student accountable if you don't understand the reasoning involved in charging them with plagiarism? What kind of evidence would you need and how sufficient should that evidence be before you take the word of an algorithm against someone else? As discussed, algorithms are increasingly being used to detect everything from recidivism and good potential job hires to cancer, and to aid in medical diagnoses. Significant ethical

---

1   Marche 2022.

consequences will follow from the use of these algorithms, even though we may not always understand how they work.

If a process isn't transparent, we call it 'opaque'. **Opacity** in AI refers to the ways in which an algorithm lacks transparency; that is, we can't explain how it works or why it reached a conclusion. Lack of transparency in any system can create ethical problems, especially when we also consider the possibility of error. As discussed in the previous chapter, an algorithm can be subject to false positives and false negatives. Given the possibility that an algorithm can be wrong, how much trust should we put in its answers if we can't explain how it reaches those answers?

This chapter examines issues involving the ethics of belief by considering the following questions:

1. Is it ethically acceptable to judge, evaluate, or appraise people if they don't know the basis for these judgements?
2. Is it ethically acceptable for you to accept the conclusion of an algorithm if you don't know how it reached that conclusion?
3. Does opacity within an algorithm prevent us from being ethically accountable for the possibility of error?
4. Given the risks of potential error from the use of AI, how strong should the evidence be to justify its usage, particularly if the workings of the algorithm are not (and may not be) fully known?

We will need to consider the nature of opacity in AI and what our ethical responsibilities are when it comes to the sufficiency of evidence for our beliefs. To grasp these ethical considerations, we will examine William Kingdon Clifford's "The Ethics of Belief" and Heather Douglas's ideas about inductive risk. To better understand the ethical consequences at stake, the chapter will review several case studies, including employee evaluations, hiring algorithms, and medical diagnoses.

1. Opacity and Its Ethical Significance

In 2011 a teacher in Washington DC named Sarah Wysocki was fired following a performance evaluation. Wysocki had been working at the school for two years and had received positive reviews from parents and peers. Her Assistant Principal wrote of Wysocki that "it is a pleasure to visit a classroom in which the elements of sound teaching, motivated students, and a positive learning environment are so effectively combined."[2] This was only two months before she was fired. So, what happened? Wysocki was subject to an evaluation system known as IMPACT, which counted for 50% of her appraisal.

IMPACT uses an algorithm to determine a teacher's skills at teaching math and language by trying to determine how much of an advance or decline in student performance could be attributed to the teacher. The idea was to weed out bad teachers by firing those with scores in the bottom 5%. As her students fell short of statistical predictions, Wysocki received a low score. But how do we know how much of a student's performance is owing to the teacher? The algorithm attempts to control for other factors that might affect student performance, such as poverty, but it's impossible to predict all external or environmental factors that could affect student performance.

Students in Wysocki's class were compared with statistically similar students across the city to detect differences in learning performance. Given all of this, it could be difficult to explain a specific result. As Wysocki later said, "I don't think anyone understood them."[3] She believed that the lower-than-expected student performance was owing to many in her class having inflated grades from the previous year. Over half of her class had previously come from a different school with a record of unusually high numbers of answer sheet erasures with wrong answers changed to right.

Wysocki did not understand how she was being evaluated and ultimately what her evaluation really meant. This process was opaque to her despite being subject to the algorithm's decisions, and it was also likely opaque in several ways even to the developers of the algorithm. While they lack full transparency, algorithms can also produce erroneous results. Everyone who is evaluated in such ways, from patients relying on medical test results to someone applying for a loan, is subject to a system lacking in transparency and yet subject to error. The fact that an algorithm can produce erroneous results that will result in real world consequences for those who must live with its decisions raises significant moral questions.

---

2   Turque 2012.
3   O'Neil 2016, 5.

1.1 WHY OPACITY?

Opacity in AI can take many forms and can occur for many reasons. For example, machine learning, taking inspiration from the animal brain, can employ **artificial neural networks** (ANNs), which can simulate the way brains learn by using layers of neurons connected in such a way as to permit the transmission of information for the purposes of problem solving. Recall from Chapter 3 that machine learning seeks to discover a function (f) that determines how changes in independent variables in the input data (represented by X) will impact the output data (represented by Y). As Karay Karaca notes, machine learning is based "on the assumption that there exists a target function $f: X \rightarrow Y$ that correctly maps all the given inputs to the corresponding outputs for a given task."[4] ANNs allow for an automated process to discover that function.

In an artificial neural network, each "neuron" holds a number. Several layers of neurons are connected to each other, allowing neurons to signal other neurons connected to it based on the strength of those connections, also known as weights. Different layers in an ANN perform different functions. For example, in an image recognition application, the first input layer might consist of neurons for each pixel in the image, with each neuron containing a number corresponding to the shade or colour of the pixel. In the final output layer, the number of neurons would correspond to the number of expected answers that a model might produce. An image classification algorithm designed to identify "dogs" and "non-dogs" as two possible outcomes would have two neurons in the output layer.

In **deep learning algorithms**, input and output layers are connected by several in-between layers of neurons called hidden layers. The expectation is that within these hidden layers, the weights of the neurons from the input layer will allow those neurons to signal neurons in the first hidden layer which might recognize simple patterns. Those neurons will then signal the next layer which might then combine those recognized patterns into a probabilistic guess about whatever is to be predicted (Figure 14).

Learning takes place using a process called back-propagation where the weights of each neuron are adjusted through an automated process

---

4   Karaca 2021, 7.

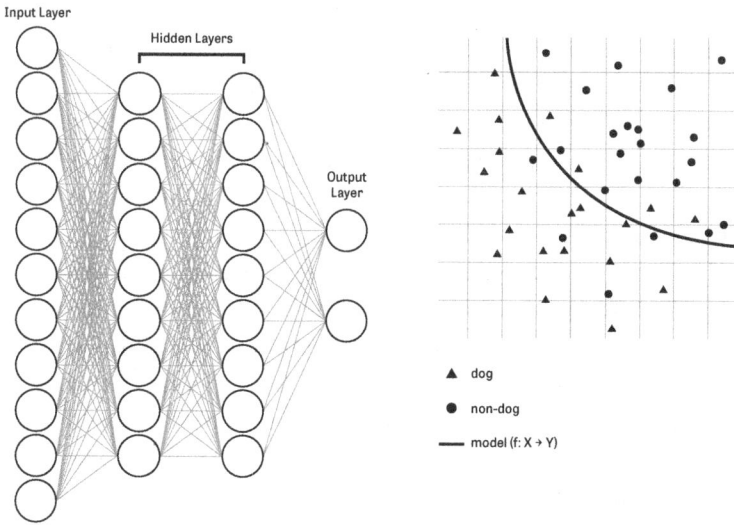

FIGURE 14 · The image on the left represents an artificial neural network. If each neuron in the input layer represents the value of a shade in a pixel of an image, the neurons on the output layer correspond to the decisions of the model for the given input. For example, the outputs might be "dogs" or "non-dogs." The result would be a model containing a function (f: X→Y), like the image on the right where a simpler function is represented by the solid black line that can distinguish between dogs represented by circles and all other things represented by triangles. Such a model would be far more complex and include several thousand dimensions.

with the result being different model curves trying to correctly separate and categorize the data until error is minimized. The problem is that it can be very difficult to determine which patterns those hidden layers are looking for or what causes them to signal other neurons. Models developed through deep learning may involve thousands of hidden layers calculating millions of points of data within a 10,000-dimensional space.[5] Once a model is trained it will produce correct answers, but the sheer amount of information involved makes it difficult to explain how it is working, making the model a **black box**. We will discuss black box opacity in section 3.1, but there are other reasons opacity exists in AI.

Opacity can be caused by social reasons as well, such as corporate trade secrecy. Like the local restaurant's "secret sauce," the propriety algorithms of tech companies might be closely guarded.[6] One study

5   Kearns and Roth 2020, 7.
6   Tracy and Wells 2023.

revealed that an entire algorithm can be stolen simply based on gradient-explanations, a process that tests how variations in input data affect the network's output.[7] By experimenting with the model in this way, one could potentially reverse engineer the model. Also, the more an algorithm is understood, the easier it is to manipulate results to obtain a preferred outcome, just as one might cheat on a standardized test by memorizing the questions.[8]

A third reason why corporations will avoid making their algorithms transparent is to avoid accountability. For example, corporations such as 3M and Dupont produce chemicals known as per- and polyfluoroalkyl substances (PFAS) for products like Teflon. These chemicals are known to contribute to high cholesterol, pregnancy-induced hypertension, testicular cancer, and kidney cancer.[9] PFAS are also known as "forever chemicals" that accumulate in the body and in the environment and remain there for a long period of time.[10] Yet, despite knowing for decades that these chemicals were accumulating in their employees' blood, and that this may have been linked to birth defects in the children of pregnant employees, Dupont did not make this information public.[11] Similarly, Facebook didn't disclose that it was selling information for housing advertising using racially motivated data in violation of the Fair Housing Act.[12]

Another reason for opacity in AI is owing to what could be called a "transparency paradox," which is when a corporation attempts to make their algorithm more transparent, exposing it to greater risk.[13] For example, the more transparent an algorithm is, the more vulnerable it can be to attacks.[14] In addition, explanations of what an algorithm is doing can be manipulated to reduce public trust in the algorithm.[15] Releasing more information about the workings of the algorithm can expose a company to greater security and liability risks, thus incentivizing corporations to be less transparent.

---

7   Milli et al. 2019.
8   Note that data manipulation is indicated in Figure 15. For further discussion of this, see Chapter 6.
9   European Environmental Agency 2019.
10   Bilott 2020.
11   Rich 2016.
12   Nix and Dwoskin 2022.
13   Burt 2019.
14   Shokri, Stroble, and Zick 2019.
15   Ribeiro, Singh, and Guestrin 2016.

1.2 THE ETHICAL SIGNIFICANCE OF OPACITY

Understanding the ethical problems of opacity is a matter of understanding the ways in which AI can be said to be opaque, with respect to both the consequences that AI will produce and in the motivations that governed its development and usage. Opacity can take on different ethical meanings depending on what is lacking in transparency and on whether we are on the receiving end of the consequences of AI or participating in its development.

We can think of machine learning as being part of a system that includes various agents interacting with the model, the input data, and those who are affected by the output choices. This is called a **machine learning ecosystem**, and it includes stakeholders such as *creators and developers* who create machine learning models, *operators* who provide input and receive outputs from the model, *executors* who carry out decisions informed by the model, *decision-subjects* (or end-users) who are affected by decisions made by executors, *data-subjects* whose personal data is used to train the model, and *regulators* who are responsible for investigating and regulating the machine learning ecosystem.[16]

According to Carlos Zednik, "opacity," "transparency," and "explanation" will mean different things to different stakeholders in the machine learning ecosystem depending on what they consider to be epistemically relevant.[17] For example, a decision-subject will likely be interested in explanations that cite reasons for a model's output. A regulator may be more interested in understanding the environmental regularities that are being tracked by a model and whether it is processing data fairly or is violating privacy rights. A creator might be more interested in explanations that allow for intervention with the model to produce specific kinds of outputs.

We can use a variation of our diagram from Chapter 3 to represent the morally salient features of the machine learning ecosystem, such as the possibility that the model might get the wrong answer, and to understand which features are more important to different stakeholders (Figure 15). While creators will be responsible for the intentions and background assumptions that go into a model, decision-subjects will be the ones experiencing the morally salient consequences of the model's output.

---

16   Tomsett et al. 2018, 9.
17   Zednik 2021, 269.

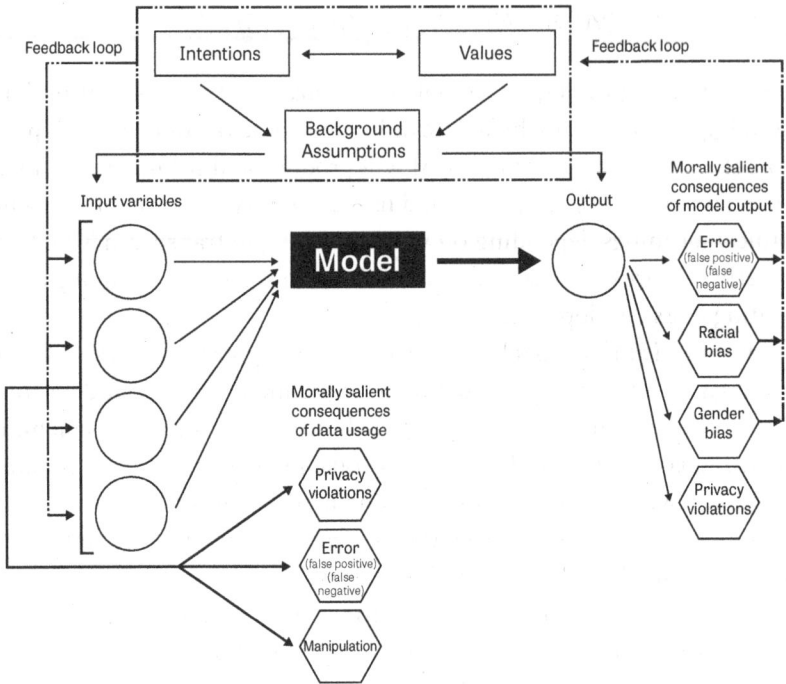

FIGURE 15 · This diagram represents the morally salient features of a machine learning ecosystem, including the morally salient consequences that follow from the collection and use of data, the assumptions that justify the usage of a model, and some of the morally salient consequences of the output of that model. While data-subjects may have to deal with the morally salient consequences of their data being collected, and decision-subjects or end-users must deal with the consequences of a model's output, creators and developers must concern themselves with the intentions and background assumptions that justify the model, and to an extent the other remaining morally salient features and consequences of the ecosystem.

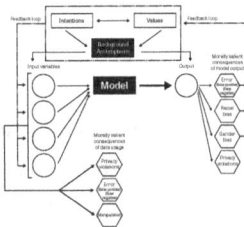

FIGURE 16 · From the perspective of creators and developers, the black box characteristics of the model make it opaque to them. Commonly held background assumptions will also not be transparent to them.

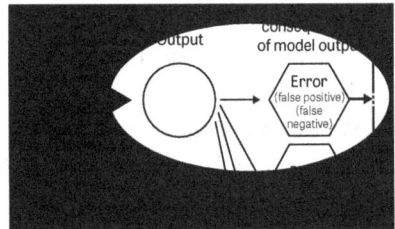

FIGURE 17 · From the perspective of decision-subjects, the only aspect of the ecosystem they will be aware of is the model's output and the morally salient consequences they may have to experience as a result. To them, almost the entirety of the ecosystem is a black box.

First, consider how the development and use of AI could be lacking in transparency. The black box problem makes the decisions of algorithms difficult to explain. However, recall from the previous chapter that a model developed from AI is built on background assumptions informed by our values and intentions. When background assumptions are shared by all members of a community, they acquire an invisibility that shields them from criticism. A background assumption may be required to justify a model developed by AI, yet it may not be apparent to the developer. Thus, commonly accepted background assumptions can represent another form of opacity (Figure 16).

Another form of opacity exists if we don't understand what data is being used for training and which variables are factored into the model. For example, if you are applying for a loan, you might not know that the algorithm is using your address as part of an e-score to gauge your reliability. The intentions of a developer might also be opaque. For example, a corporation developing a facial recognition algorithm might not disclose that its purpose is to target protestors. A teaching evaluation algorithm might be designed with the intention of busting teachers' unions. A hiring algorithm might not be trying to identify the best candidate, but to exclude as many applications as cheaply as possible.

Decision-subjects and the developers don't share the same understandings, yet the fact that the decision-subject presumably faces the brunt of the consequences of AI outputs means that opacity will affect them differently. They simply do not understand the algorithm, yet it affects them more (Figure 17). As Zednik notes, "Practically, end users are less likely to trust and cede control to machines whose workings they do not understand."[18]

The level of trust that we can reasonably expect the end-user to have will be different from that of the developers and this will produce ethical consequences. For example, if you don't know where the data used to train the AI comes from, if you don't know how representative it is, then it makes it difficult for the end-user to assess how generalizable its conclusions are. If you don't understand what the algorithm is really trying to measure, you can't even say for sure if it is really making an error. Because of this difference and because the end-user knows even less, their moral responsibilities, when it comes to responding to and

---

18   Zednik 2021, 266.

believing the findings of an AI, will differ. We will consider these ethical responsibilities from the perspective of the developer first, followed by that of the end-user. First, however, we must clarify the ethical limits of trust and belief, which in turn will clarify what ethical responsibilities we might have when we adopt beliefs based on the output of an AI that is opaque.

2. The Ethics of Erroneous Belief

How does the ethics of belief relate to ethical concerns about AI? What is a responsible belief? Do our beliefs matter to us only as individuals or also to our broader community? The ethics of belief is an umbrella term encompassing several issues at the crossroads of ethics and epistemology (the study of knowledge). It holds that we are morally obligated to believe responsibly, as others trust us to do the same.

2.1 W.K. CLIFFORD'S "THE ETHICS OF BELIEF"

The term was coined by philosopher and mathematician W.K. Clifford, who endorsed a version of **evidentialism**, summed up in his memorable statement, "It is wrong always, everywhere, and for anyone, to believe anything upon insufficient evidence."[19] Evidentialism holds that one is justified in believing something only if they have evidence that supports that belief. What does Clifford, writing during the Victorian era, have to offer so we may understand the ethics of AI? According to Francisco Uribe, "I cannot think of anyone whose ideas are more relevant for our interconnected, AI-driven, digital age."[20] Our beliefs reach across communities and even globally; indeed, Clifford thought they go further: they also impact future generations.

Clifford opens "The Ethics of Belief" with an illustration of a wrongful belief in the form of a story. Imagine that a shipowner inspects his vessel and finds it's damaged, weathered, old, and ready to be retired. However, it's scheduled to leave tomorrow, and it's supposed to carry many emigrant families across the Atlantic Ocean to a land of

---

19   Clifford 1879, 186.
20   Uribe 2018.

opportunity. Instead of listening to his doubts and calling for inspectors, the owner stifles these worries and convinces himself that the vessel will make it, reminding himself of past successful trips and placing trust in Providence (i.e., God's goodness). Still, the owner is ready to collect his insurance money if the ship sinks and to never speak of his doubts about the vessel.

While Clifford was composing "The Ethics of Belief," a sinister insurance scheme unfolded. Greedy shipowners and merchants routinely overloaded and heavily insured unsafe ships. If the vessels sank, they would gain insurance money. Worse, the ships often carried poorer emigrants, and sailors dreaded crewing these vessels. There was an ominous but apt name for them: coffin ships. Everyone from Florence Nightingale to Queen Victoria opposed this terrible practice. However, the problem was a lack of nautical regulations at the time, compounded by the fact that several of these shipowners were also powerful politicians able to avoid legal culpability. Notice that the coffin ships issue mirrors Clifford's own case study of the shipowner, including the issue of insurance money and the temptation to supress doubts for the sake of financial gain.

We can understand Clifford's project if we turn to two core points: (1) beliefs influence actions; (2) beliefs matter to inquiry. In magnifying these points, we can further consider Clifford's ethics, his evidentialism, and Uribe's observation about the interconnectedness of belief and its relevance to the ethics of AI.

Clifford appeals to an evolutionary framework in which he thinks that we, as a species, have evolved in tribes and groups, and possess an evolved conscience. Drawing on this evolutionary framework and account of moral responsibility, Clifford argues that "right is an affair of the community and must not be referred to anything else."[21] He continues: "The first principle of natural ethics, then, is the sole and supreme allegiance of conscience and the community," extolling the virtues of a mutually trusting society. If we break the bond between conscience and the community, we will engage in wrongdoing or be liable for doing so.

What does Clifford mean by something being morally wrong? He explains: "In general it is wrong to injure [any member of the community]

---

21   Clifford 1879, 171.

in any way in our private capacity and for our own sake."[22] Stressing the interconnectedness of morality, moral wrongs are a deviation from the community, primarily through betrayal.

With this outline of Clifford's ethics, we can see why he places such weight on our acting responsibly and for the community's sake. Do we need an ethics of *belief*? Maybe you think that action is one thing, something we do, whereas beliefs are different: we don't necessarily act on our beliefs. However, in "The Ethics of Belief," Clifford argues that "For it is not possible so to sever the belief from the action it suggests as to condemn the one without condemning the other." Clifford explains: "If a belief is not realized immediately in open deeds, it is stored up for the guidance of the future."[23] Beliefs prepare us to act on them.

To see this connection between belief and action, let's say, you believe that it will rain today. Suppose you acquired your belief from the weather report, a reliable source. You peer out your window and see storm clouds. If you believe it will rain and want to stay dry, you are ready to bring an umbrella, take the bus, or perhaps cancel your plans. Clifford reminds us of how beliefs make us apt to act on them, and when we have prejudiced or biased beliefs, they can have drastically wrong and sometimes catastrophic results. For example, some US citizens believed the 2020 presidential election was stolen due to rigged voting machines. The belief that the election was stolen made such citizens ready to act drastically and explained why they would, for example, storm the Capitol. How we form our beliefs, then, matters to whether we will act responsibly.

At this juncture, we can return to the case concerning AI technologies that we sketched at the beginning of this chapter: the teacher who suspects students of using ChatGPT to cheat on term papers. Suppose the teacher believes a particular AI program is a reliable plagiarism detector. The teacher will likely use it to uncover student cheating. Again, our beliefs make us prepared to act on them. The example reinforces the need to form our beliefs responsibly and preserve the social trust that Clifford's ethics underscores. A teacher who implements the AI cheater detector here will do so again if they continue to believe it works, establishing a habit or practice. One can imagine how social trust

---

22   Clifford 1879, 174.
23   Clifford 1879, 181.

would be broken if the teacher arrived at their belief by stifling their doubts about the reliability of the algorithm if problems arise. Clifford stresses the need for inquiry, considering careful investigation the best means to arrive at our most responsible beliefs, claiming that "every time we let ourselves believe for unworthy reasons, we weaken our powers of self-control, of doubting, of judicially and fairly weighing evidence."[24] This leads to what Clifford calls the duty of inquiry.

How many doubts must we raise, how many questions must we ask, and how much evidence must we collect to give us a right to believe? These questions are germane to Clifford's project, even if he does not offer precise answers. Doubts, questions, and evidence are the ingredients of inquiry. Clifford maintains that the shipowner "had no right to believe on such evidence as was before him. He had acquired his belief not by honestly earning it in patient investigation, but by stifling his doubts."[25] Consider if the vessel had made it, managing to transport all the families across the Atlantic to the new world. Should we still consider the shipowner blameworthy? Yes, Clifford insists. The *origin* of the belief" is salient and shows that the shipowner should have cancelled; it doesn't matter how things turned out.

Indeed, the shipowner would still have *insufficient* evidence for the belief that the ship was seaworthy, as he failed to follow up on his initial doubts concerning the vessel being old, weathered, and rickety. He would be acting in his private interests, not for the community; and his doubts, questions, and evidence show that he lacked a right to believe in its safe passage. The shipowner was reckless and negligent. He had a duty to inquire; he should have followed up on his doubts and questions instead of suppressing them.

There are two points to consider. First, beliefs tend to generalize. If one believes something—say, that an election was stolen or that an AI program can reliably detect cheating—one will be apt to accept similar beliefs or reason from those beliefs (e.g., be ready to protest the election or accuse students of plagiarism based on the program). Second, once we believe something, it isn't easy to modify or update that belief, even when faced with solid countervailing evidence. Our beliefs can persevere even if false and prevent us from being fair inquirers, underscoring

---

24   Clifford 1879, 185.
25   Clifford 1879, 178.

how significant it is that we fulfill the duty of inquiry before forming beliefs.

Continually failing to meet the duty of inquiry will plunge us into credulity. We will "lose the habit of testing things and inquiring into them." We will diminish our powers of inquiry and our practices of carefully examining what we doubt or question. Clifford contends that if this were to happen, we could become almost like naïve children who automatically believe what they are told—the existence of Santa Claus or the Tooth Fairy, and so on. Such would be an enormous betrayal to future generations who rely on us, to a significant extent, to uphold standards of inquiry.

The point parallels Clifford's natural ethics. Suppose we avoid our responsibilities when it comes to belief and never seek to improve our thinking. This resembles how a muscle atrophies or shrinks if not used. In this case, we would commit a severe injustice to the future community by eroding its abilities and denying it the natural resources to remain moral. Improvement never comes from neglect but from responsible engagement.

What exactly does Clifford mean when he says, "To sum up: it is wrong always, everywhere, and for anyone, to believe anything upon insufficient evidence"? Clifford never clarifies what an adequate evidential bar would be or how much evidence is enough. Instead, he says: "Inquiry into the evidence of a doctrine is not to be made once for all, and then taken as finally settled. It is never lawful to stifle a doubt; for either it can be honestly answered by means of the inquiry already made, or else it proves that the inquiry was not complete."[26] Suppose we want to believe something, despite our doubts, for our private interests. We know this blatantly violates the duty of inquiry and is morally wrong. It betrays the community. It breaks social trust, including future generations' trust in us. That may sound odd but think of your trust in previous generations in not letting you inherit a world rife with toxic prejudice and foolish superstition. Maybe you think that has already happened, that past generations failed us. But then you are affirming Clifford's call for an ethics of belief.

As we enter the AI age, we face many significant issues which challenge us as careful inquirers and responsible believers. Although

---

26   Clifford 1879, 187.

regulating AI may help, we will all have to do our part. Suppose we are responsible for implementing AI safely. In that case, we must weigh the risks, including the doubts, questions, and evidence we and others in the field have, as the relevant technology will alter our belief-forming practices now and in the future. However, even if we are on the other end, as users or consumers of the technology and its services, we must also be vigilant. Suppose we invest our trust too readily in these novel technologies, simply defaulting to whatever they tell us. We may weaken our investigative powers and ability to fulfill the duty of inquiry. AI may make us credulous and stubborn in our existing beliefs or prejudices. This is why Uribe argues that Clifford's ideas have never been more relevant in this AI era. Conducting ourselves responsibly now is critical, given the influence current practices will have on successive generations.

2.2 INDUCTIVE RISK

In Chapter 2 we discussed inductive risk as it pertains to the moral responsibilities of scientists. To recap, as explained by Richard Rudner, the concept holds that, "our decision regarding the evidence and respecting how strong is 'strong enough', is going to be a function of the importance, in the typically ethical sense, of making a mistake in accepting or rejecting a hypothesis ... How sure we need to be before we accept a hypothesis will depend on how serious a mistake would be."[27] So in essence, inductive risk asks us to consider how strong or how much evidence we should have before we claim that something is true relative to the ethical risks of getting it wrong. We know that in machine learning, the aim is to find a hypothetical model that will yield correct outputs for certain inputs, meaning that any potential model is also susceptible to making mistakes that produce ethically significant consequences.

According to Heather Douglas, scientists (and by extension AI developers) are ethically responsible for inductive risk concerns in so far as they are responsible for not being reckless or negligent in producing such a model and then using that model in the world on decision-subjects. In Chapter 3 we discussed how a developer might consider different trade-offs between trying to prevent false positives and false negatives in a model.

---

27   Rudner 1953, 2.

For example, if we want to test a new chemical to see if its use is safe, and we are concerned about the health impact if we mistakenly say that the chemical is safe when it isn't, we might demand a higher standard of evidence to prove that the chemical is safe and thus increase the risks of a false positive for unsafe factors while decreasing our risk of false negatives for unsafe factors. If, on the other hand, we were less concerned about the public health risk, as this chemical would be greatly economically beneficial, we might be more inclined to try and certify it as safe by increasing the risk of a false negative and decreasing the risk of a false positive that the chemical is unsafe.

A false positive is known as a type I error while a false negative is known as a type II error. Thus, the heart of inductive risk thinking is considering how much evidence you should demand given the ethical consequences of a type I or type II error. Like Clifford's ethics of belief, arguments regarding inductive risk demand that we have an ethically justified amount of evidence to support whatever we are prepared to say is true.

So far, we have not considered how inductive risk can be applied to the fundamentals of accepting a hypothetical model. When a scientist decides to accept a hypothesis, they compare this against what is called the **null hypothesis**, which is the idea that there is no statistically significant relationship between the variables being measured in relation to each other. To claim that there is a statistically significant relationship between the variables hypothesized, the scientist must reject the null hypothesis.

When dealing with data in statistics, however, we don't always see what we might expect. If I have a fair coin then statistically flip it, I should have an equal chance of getting heads and tails. However, it is always possible that I could flip a fair coin a hundred times and get heads each time. The result of repeated tests of flipping a coin a hundred times will be a bell curve that includes those occasions where we obtained statistically possible yet unlikely results of a large number of heads or of tails (Figure 18). When scientists collect empirical data to test their hypotheses, they face a similar situation whereby if a given hypothesis were true, there might be results we would normally expect. However, we might also get results that are statistically possible, but just very unlikely. How can we tell if the null hypothesis is false, or if we might see statistically possible yet unlikely results?

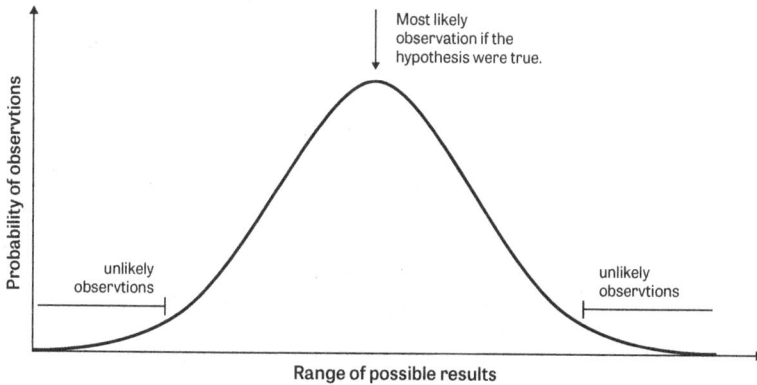

FIGURE 18 · When we take empirical samples of the world, the results can be expressed by a bell curve, where there are very statistically likely observations but also statistically very unlikely observations; just as flipping a fair coin a hundred times could result in a hundred heads, there are results that are statistically possible but just not statistically likely. Suppose you were worried that the coin wasn't fair. How different would the results need to be from what would be unlikely observations if the coin were fair before you would conclude that the coin isn't fair?

To determine if they can reject the null hypothesis, scientists need to compare the results they obtained to the results they would expect to obtain if the null hypothesis were true. The problem is that there can be an overlap between these two expected results whereby, for example, we might obtain a result that is consistent with our hypothesis, but this could also be consistent with a very unlikely observation if the null hypothesis were true. Scientists use a cutoff point known as a p-value to determine if the result they are seeing counts as a statistically significant enough difference from the null hypothesis to justify rejecting the null hypothesis.

The p-value test is a function of whether the given observed result lies outside of the boundaries of the null hypothesis relative to how dispersed that data is. Typically, this is done by calculating the standard deviation of the null hypothesis, or how much the data deviates from the mean. One standard deviation, for example, represents 68.2% of the distribution, while two standard deviations capture about 95% of the distribution (Figure 19). For example, imagine that we examine exam test scores and calculate that the mean (average) was 78% with the standard deviation being 11.33. This means that within one standard deviation, the scores were 78 +/– 11.33, in other words between 66.67% and

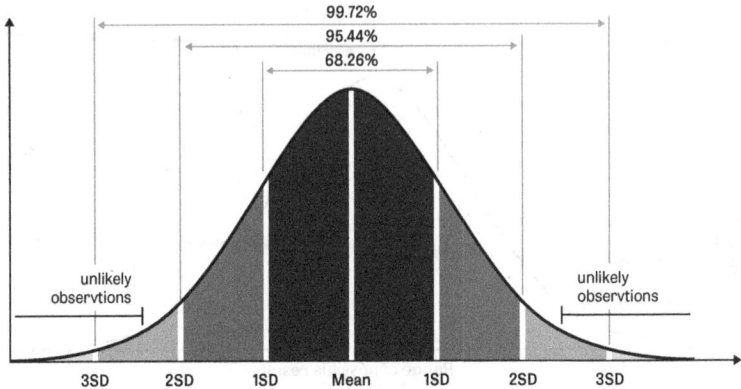

FIGURE 19 · In a normal distribution, each unit of standard deviation from the mean is a degree to which we can confidently generalize about the hypothesis. The greater the standard deviations, the greater degree of statistical confidence we can generalize over; however, the distribution of the results gets larger. The more confident we try to be, the greater variation in data we must contend with. The normal standard of a statistically significant result is two standard deviations representing 95% confidence.

89.33%. Another way of understanding this is to say that these deviations capture how statistically confident we can be about what results we would expect to find within a certain range.

In other words, we can use standard deviations to express how confident we can be that certain results would be true within certain ranges. Typically, scientists are looking for a p-value result of at least less than 0.05, which means that we can be at least 95% confident that our actual result does not overlap with what we would expect to see if the null hypothesis were true. So, if our actual result is outside of the range of two standard deviations of results that we would expect to see if the null hypothesis were true, we can conclude that the null hypothesis is false with 95% confidence (Figure 20). This means that there is a 5% chance that the null hypothesis could be true, and we are just seeing statistically unlikely results.

Notice the risky trade-off. Your results could be potentially consistent with your hypothesis, or they could potentially be consistent with what we would expect to see in very unlikely cases if the null hypothesis were true. How confident should we be that our results aren't consistent with the null hypothesis? The typical norm to publish something as a **statistically significant result** is 95% confidence. However, in some

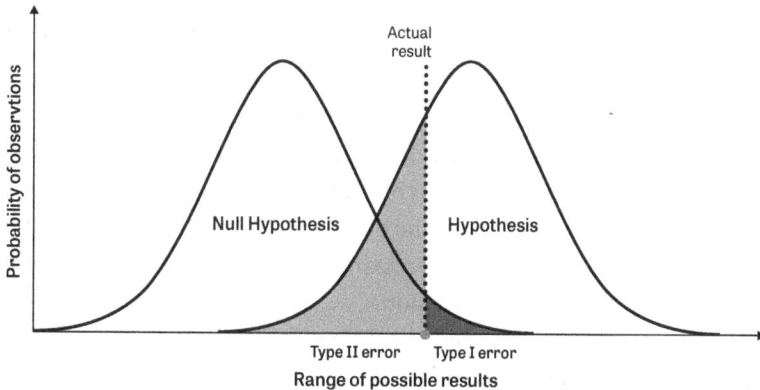

FIGURE 20 · In a situation like this, if we set our desired p-value higher (higher than 0.05 or less than 95% confidence), we increase the likelihood of creating a type I error, or a false positive result, by concluding that the null hypothesis is false when it isn't. If the null hypothesis is false, we decrease our risk of a false negative. On the other hand, if we demand a higher standard of evidence by expecting a lower p-value (lower than 0.05 or greater than 95% confidence), we decrease the risks of a false positive but increase the risks of a false negative if the null hypothesis is false.

cases the finding might be so important and the risk of getting it wrong so great that we might demand a higher standard of evidence. For example, we could demand a level of statistical confidence beyond three standard deviations. That would equal 99.7% and require a p-value of less than 0.003.

During the discovery of the Higgs boson particle, scientists sought an incredibly high standard to prove their results. Their results were significant beyond 5 standard deviations, or roughly equivalent to 99.99997% confidence level that they would not be getting the results that they were getting if the null hypothesis were true. Members working at the Large Hadron Collider felt that the intense public spotlight "made it clear that there was an opportunity to try to show science of a very high quality to the general public in an environment where there was skepticism about some scientific claims. Certainly, making a discovery announcement that subsequently turned out to be erroneous carried a very high cost."[28]

In other words, partly due to social pressure, the scientists were more concerned about the risk of committing a type I error, so they demanded a very high standard of evidence. Thus, we can now appreciate how

28   Staley 2017, 51.

scientists can manage inductive risk concerns at the statistical level. If a scientist is concerned about the ethical consequences of reaching an erroneous conclusion, they can adjust their p-value to seek a stronger or weaker degree of evidence. If we have minor concerns about the ethical risks of getting it wrong and the discovery would hold great promise, we might demand a high standard to prove that our conclusion is unsafe. If we were more concerned about the possible harmful consequences if we are wrong and we were testing for harm, we might adopt a low standard of evidence to reach that conclusion.

Given what we have learned, we do have certain ethical responsibilities for what we claim is true. We have a duty to make sure, for example, that if we are going to use a statistical model, it must be the product of a process of inquiry that does not stifle doubt, and we must be prepared to consider the consequences if that inquiry ends in error. Given these ethical responsibilities to make sure that we have sufficient evidence for our beliefs, what are our responsibilities given the opacity encountered in AI? As we consider issues relating to the opacity of models derived through an automated machine learning process and the possibility that they might produce erroneous results, this raises questions about how an AI developer can manage inductive risk concerns, particularly if they don't understand the model.

INQUIRER'S TOOLBOX

13. Do I have sufficient evidence to support the beliefs that I have about this moral case?
14. Given the risks posed by getting it wrong, is a higher or lower standard of evidence justified?

In the meantime, given that we are all responsible for not acting recklessly or negligently and assuming we ought to have sufficient evidence for our beliefs, we can add these insights to our inquirer's toolbox.

3. Opacity from the Perspective of Creators, Developers, and Executors

3.1 BLACK BOX OPACITY

As noted in Section 1.1, machine learning that deals with complex domains featuring complex datasets can make use of deep learning using ANNs. Multilayer perceptrons, convolutional neural networks, and recurrent neural networks are examples of machine learning that make use of such ANNs, although each is structured and functions in different ways.[29] Neural networks can approximate anything that can be expressed as a function. The idea is to use training data to discover a function that best separates or categorizes that data according to the qualities that we are interested in. But why is the model an opaque black box? Why can't we know what is going on in the hidden layers of a neural network?

As discussed, ANNs use a collection of artificial neurons arranged in layers with weighted connections to each other that allow for the transformation of input data into a desired output. Each neuron in the network holds a number (or more accurately a function that produces a number), but since not all inputs are equally relevant to finding a solution, the input data provided at the input layer must be multiplied by a corresponding weight parameter.[30] A neuron in the input layer will activate a neuron in the hidden layer based on a summation of all data from the input layer according to these relative weights plus a number called a bias using the formula $f(\sum_{i=1}^{n} W_i X_i + B)$.

If the weights and biases are assigned randomly in a neural network, you will get random answers, as the strength of each connection of one neuron to another is random. To arrive at a model that articulates a suitable function, those weights and biases will need to be adjusted until the network produces more accurate results. This is where training data becomes important. Let's consider an example of a neural network being trained using supervised learning. In this case the **training data** is labelled, meaning that if the network produced an error, this can be evaluated relative to what the right answer should have been. To arrive at the desired function, we want to expose our neural network to the training data and allow the neurons to pick up on statistical patterns in the training data to enable a correct prediction. The idea is to automatically adjust those weights and biases of the neurons to minimize the overall error of the model based on the patterns detected in the training data. This is tested by seeing how well the model generalizes to data it was not previously trained with using **test data**.

---

29   Karaca 2021, 20.
30   Dignum 2020, 28.

Watch a video explaining machine learning and modelling.

To make adjustments to the various weights and biases, developers utilize a mathematical formula called a **cost function** to measure the difference between the desired outcome determined by the answers in the training data and the actual outcome produced. A higher number will indicate a higher degree of error, while a smaller number a lower degree of error. The developer can average the degree of error for every training example to indicate a total overall "cost" or error for a given state of the neural network. In other words, it is a measure to determine how well a machine learning (hypothetical) model performs for a given dataset.

The machine "learns" when it can adjust those hypothetical models by changing those weights and biases to produce a model that will reduce that cost (or overall error). This is where overfitting and underfitting become relevant. If there is too little data, the model will not be sophisticated enough to find the appropriate function. This will result in errors and is called **underfitting**. Alternatively, a model can mirror the training data too much, whereby it can make accurate predictions using training data but can't generalize to new data because the model fits the old training data too closely. This is called **overfitting**. So, if an attempt is made to find a model with zero errors the result will likely be overfitting (Figure 21). Thus, the goal is not to eliminate all possible error, but to minimize error so that it still generalizes to test data without overfitting.[31]

To discover the correct model, an automated process takes place using an optimization algorithm called **gradient descent** where each of the weights and biases are adjusted to determine what changes to those weights and biases are most efficient at minimizing the cost or overall network error. Imagine that you are a hiker and you want to descend a mountain into a valley. The question is this: What step you could take that would be the most efficient to get you to the bottom the fastest? As you plan a step in one direction or another, the slope of the descent changes. Ideally, we want to move in the directions where the slope is the steepest so that each step moves you further to the bottom with the fewest steps. Gradient descent works in a similar manner. Each possible

---

31   Karaca 2021, 11.

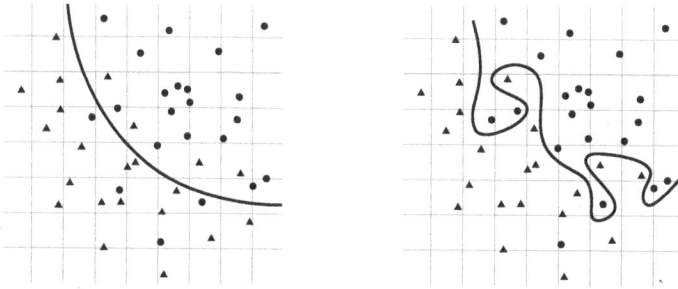

FIGURE 21 · The model on the left will produce some errors but will likely generalize better to real world data than the model on the right, which will produce zero errors because it has overfitted to the training data.

weight and bias value would result in a specific error cost. If we were to graph the possible directions we could move in by changing those weights and the corresponding error rate for each, we will arrive at something like Figure 22. If we adjust those weights and biases, it would be like taking a step down the mountain. Just as stepping in certain directions will result in a steeper slope down the mountain and others will be less steep, gradient descent seeks to find which changes (what steps) in the weights and biases of each neuron will result in the steepest descent towards a valley where error is minimized, or what is called the local minimum.

Gradient descent works a bit like the hiker in that it seeks to find which steps will be the most efficient at finding the steepest slope to the bottom. The higher the gradient, the steeper the slope, and the faster a network will reach the local minimum (in a word, learn). However, as the minimum approaches, the slope will be less severe, so it must adjust each step it takes as it goes in order to avoid the risk of overshooting the minimum by taking too large a step. In complex neural networks, there can be tens of thousands of different dimensions of weights and biases that produce the overall model. So, each possible variation of those tens of thousands of variations will produce a different overall cost function error value.

Even if a model is discovered that produces few errors in training data, the real test is to see how well it can generalize to the world beyond the limits of training data. Thus, the model will be tested using test data

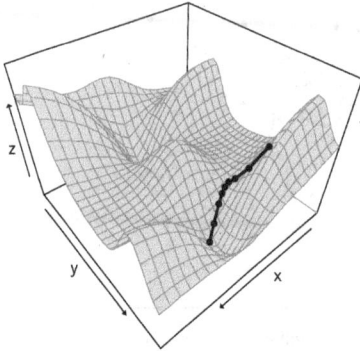

FIGURE 22 · In gradient descent, the algorithm takes repeated steps of changing the weights and biases of the neural network to produce different models with different error rates. When the error rates are graphed as a function, we can see various gradient slopes including several valleys that would represent states where the model has minimized error. We can take different steps relative to a gradient of that function (represented by the slope of the peaks and valleys) and then descend that gradient slope to find a local minimum error rate (represented by the valleys).

to determine how accurately the model can yield predictions in cases it hasn't previously been exposed to.

This discussion of AI and black box opacity is obviously technical, so let us recap the important ethical considerations. First, AI works according to the principles of statistical modelling. In other words, the thing that we want the network to do must be able to be represented in some way (using available background assumptions) as a function. Whatever function is eventually discovered will be the result of whatever statistical relationships that stand out in the training data that contribute to finding the correct answer.

Second, this process tells us that the developer might have a good understanding of the algorithm that was used to find the decision-making model, but not the model itself.[32] They have produced a model that results in few errors in making predictions using data that the algorithm wasn't trained with, but what statistical patterns those hidden layers of neurons are looking for are opaque. Optimizing the model "looks like a black box process in that humans can access the initial inputs and final outputs, but not the inputs and outputs of the modes in the hidden layers."[33] This explains why the model in Figure 16 is a black box.

Hence we must consider how this process fits into our ethical responsibilities when it comes to inquiry and examine the consequences of error. As one paper explains, "When using machine learning for medical diagnosis or terrorism detection, for example, predictions cannot be acted upon on blind faith, as the consequences may be catastrophic.

---

32   Kearns and Roth 2020, 11.
33   Karaca 2021, 21.

Apart from trusting individual predictions, there is also a need to evaluate the model as a whole before deploying it 'in the wild.'"[34]

What if the model produces erroneous results yet we don't understand the model? Who is responsible for the consequences of error that a model produces and to what degree? Imagine, for example, that due to ethical concerns we wanted to strongly avoid the risk of a false positive by increasing our odds of a false negative. How could we build inductive risk concerns into the production of a model? In the remainder of this section, we will consider these ethical questions.

3.2 INDUCTIVE RISK AND OPACITY

How can an AI developer build inductive risk concerns into a potential model and to what degree? According to Karay Karaca, the accuracy of the prediction of a machine learning model (ML) depends on how well these models generalize to new data sets; thus, "the application of ML models to big data is based on inductive generalization, and as a result, their predictions about new data sets are always prone to error."[35] Further, this establishes a clear ethical decision for the developer. Accepting a model entails accepting its empirical consequences; in other words, the inductive risk taken in accepting a particular model amounts to the risk of accepting its empirical consequences that can be wrong.[36] But how do inductive risk concerns manifest in the construction and adoption of a ML-derived model?

Karaca relies on the distinction between test error and training error. Training errors take place during the model construction process and represent errors that are made by the models using the training data which are minimized before a model might be tested against test data. Test error, therefore, represents the degree of error that results when the model is exposed to test data. Thus, test error, "can be regarded as a measure of the inductive risk associated with the use or application of a model," whereas "training error indicates the extent of inductive risk built into the model during its construction."[37] Unlike test error, training error can be directly controlled by the developer given that they control

---

34   Ribeiro, Singh and Guestrin 2016, 1135.
35   Karaca 2021, 2.
36   Karaca 2021, 6.
37   Karaca 2021, 11–12.

the process of minimizing error in the model without overfitting. Thus, minimizing training error during ML optimization serves to minimize the inductive risk built into the ML model.[38]

Karaca distinguishes between **cost-insensitive machine learning** optimization and **cost-sensitive machine learning** optimization. Machine learning can be said to be cost-insensitive if the optimization is carried out using equal costs for different types of training errors and can be said to be cost-sensitive if the optimization uses different costs of different types of training errors. Because cost-sensitive machine learning would allow us to assign different costs to a false positive or a false negative training error, we can use that to build in inductive risk considerations.

This highlights an important difference. Cost-insensitive machine learning maximizes the model's overall predictive accuracy. Alternatively, because a cost-sensitive machine could assign different costs for different kinds of mistakes (perhaps hoping to avoid worse mistakes by creating lesser mistakes), it does not aim at maximizing predictive accuracy.[39] This assumes that the results may not be fully trustworthy, and we are worried about making certain kinds of mistakes. Karaca provides the example of a bank that might wish to develop an algorithm to determine if someone is too risky to be granted a loan. The bank could use cost-sensitive machine learning to create an algorithm depending on how risk averse the applicant is.

Karaca explains: "If the bank values financial risk taking more than financial security, it will be more willing to accept applications than to reject them. This means that the cost of misclassifying (rejecting) a low-risk customer is more than the cost of misclassifying (accepting) a high-risk customer."[40] On the other hand, if the bank values financial security over financial risk-taking, it will be more willing to reject applications than to accept them. The cost of mistakenly accepting a high-risk customer would be more than the cost of mistakenly rejecting a low-risk customer. Thus, a developer could build inductive risk considerations into a model by building cost-sensitive machine learning into the model according to whether one is more concerned about the ethical consequences of a false positive or a false negative.

---

38   Karaca 2021, 12.
39   Karaca 2021, 13.
40   Karaca 2021, 13.

Is this a sufficient answer? According to Karaca, we can build inductive risk into the training process by using cost-sensitive machine learning. There is an additional inductive risk associated with deciding to accept if the model can generalize to new data using test data. But here we also only have predictive accuracy in terms of overall test error as a metric to go by. In other words, we must bet on something we don't understand with the understanding that statistically it will test correct most of the time. Karaca's framework doesn't specifically address the issue of opacity and its relationship to the inductive risk that exists within these models. So, is this the extent to which we can consider inductive risk? To see why this issue is more complicated, we can turn to issues involving statistical modelling in the sciences.

3.3 P-HACKING AND THE REPRODUCIBILITY CRISIS

In the sciences, researchers are expected to regularly publish their findings. However, most scientific journals and publications will only publish a submission reporting a finding that is statistically significant. Studies that don't find statistically significant results aren't usually published, and studies that attempt to replicate the findings of previous studies to confirm their results are rarely funded or published.[41] This can create a problematic culture with the way science is done. Scientists are rewarded when they find statistically significant results; this leads to an over-emphasis on p-values. As one young scientist said, "I feel torn between asking questions that I know will lead to statistical significance and asking questions that matter."[42]

A scientific culture that emphasizes p-values in research is vulnerable to a process known as p-hacking. P-hacking is a process whereby a developer can repeatedly perform the same experiment or repeatedly run different statistical tests to discover patterns that count as statistically significant, yet in reality there is no relationship.[43] Within a given dataset, one can find all sorts of statistically significant yet meaningless results. For example, using a dataset containing large numbers of variables from food frequency questionnaires, a study by the website FiveThirtyEight was able to find statistically significant correlations

---

41   Martin and Clarke 2017.
42   Resnick 2019.
43   Head et al. 2015, 1–2.

between drinking coffee and cat ownership, between eating shellfish and right-handedness, and between using table salt and having a positive relationship with one's internet service provider.[44]

Relying on the value of p-values as a measure of quality combined with the effects of p-hacking has resulted in a reproducibility crisis in science, where a significant portion of published works that contain statistically significant results are not replicable. A 2015 paper attempted to replicate 100 findings published in a psychological journal and only 39% of them passed.[45] In economics, the rate of reproducible studies is closer to 60%.[46] The principle of replication is central to the scientific method, which seeks to identify reliable relationships in the world, so failure to reproduce results is a problem. These problems are so endemic that meta-science researcher John Ioannidis has concluded that most published research findings are false.[47]

In response, scientists have debated the merits of changing standards for what counts as a publishable result. Some suggest that instead of using the standard of .05 (indicating 95% confidence in rejecting the null hypothesis), the standard should be 0.005 (99.5% confidence).[48] This proposal has problems because while it would reduce the number of false positive results being published, it would increase false negative results and therefore potentially stifle scientific progress. Still others suggest that the significance of p-values as a metric of quality should be reduced. A p-value does not indicate if there is strong evidence for a hypothesis; it only indicates how surprising the results would be if the null hypothesis were true. As a result, some journals have stopped publishing p-values.[49]

To what extent is AI research susceptible to the same problems? P-hacking extends to machine learning research as well. MIT Professor Sandy Pentland indicates that "according to some estimates, three-quarters of published scientific papers in the field of machine learning are bunk."[50] But what about machine learning itself? With p-hacking we've learned that through manipulation of a given dataset, one can find all sorts of correlations between variables that count as statistically

---

44   Aschwanden 2016.
45   Open Science Collaboration 2015.
46   Bohannon 2016.
47   Ioannidis 2005.
48   Resnick 2017.
49   Aschwanden 2015.
50   Kearns and Roth 2020, 145–46.

significant, and thus seem generalizable, yet in reality mean very little. The process of gradient descent essentially involves repeatedly varying the parameters and effectively trying out different models until it detects correlations of variables that allow for predictions that perform better than random chance. Thus, it is entirely possible that machine learning is susceptible to p-hacking.

An example provided by Michael Kearns and Aaron Roth looks for correlations in people between their decision to purchase a particular book and numerous other variables such as whether their car has a fuel door on the left or the right or whether their birthday falls between January and June. For this dataset, however, all these answers are effectively randomized and the decision to purchase the book or not was done according to a coin flip.

Even in a random dataset like this, certain features might be positively correlated with book purchases and some features will be negatively correlated. For example, "perhaps among [500] people with an odd number of letters in their last name, 273 bought the book."[51] Using these various detected correlations in combination we could create a model to predict if someone is more or less likely to purchase the book. As Kearns and Roth note, "this sensible-sounding methodology can quickly lead us astray. If you try this out with enough features, you'll find that you appear to get a classifier that approaches perfect accuracy—able to predict with near certainty who will buy the book—if you measure the performance of your classifier on the same data set you trained it on. Of course, we won't do better than random guessing—because customers are just flipping coins."[52]

In other words, machine learning is susceptible to the same kinds of problems that scientists face when it comes to p-hacking. A model will make predictions according to the correlations detected by the machine learning process, but while those correlations might be found in the training data, they might also be effectively meaningless. Despite this, however, the model will seem accurate when tested against training data. But why is this so ethically significant? The reason is the opacity involved in these black box models. The risk is that a machine learning process might engage in p-hacking-like behaviour, identifying

---

51   Kearns and Roth 2020, 153.
52   Kearns and Roth 2020, 153.

meaningless patterns and correlations that appear to generalize well, yet because of the black box nature of the model, the developer wouldn't be aware of this.

3.4 INDUCTIVE RISK AND THE BLACK BOX

Let's recap. A machine learning developer can follow a process whereby they have a good understanding of the algorithm that was used to find the model, but not the model itself. We understand the inputs and the outputs of the model, but we don't understand what patterns the model is relying on to yield accurate predictions. Despite this, the process that finds these models might pick up on correlations that appear to yield accurate predictions yet mean nothing. The developer can appraise the model in terms of its predictive accuracy using training data and later test data, but they don't know which patterns the model is specifically looking for. This raises a distinctive ethical problem concerning inductive risk in the face of opacity.

As Kearns and Roth note, "Algorithms—especially models derived directly from data via machine learning—are different. They are different both because we allow them a significant amount of agency to make decisions without human intervention and because they are often so complex and opaque that even their designers cannot anticipate how they will behave in many situations."[53] This is especially the case when the input data is complex and the space of possible models is very large.

This brings us to the central problem. A machine-learning-derived model might rely on correlations that seem statistically significant to derive seemingly accurate predictions, yet these correlations are meaningless. Despite this, we are relatively unaware of what correlations the model considers significant, or how significant each correlation is for the model in terms of explaining its prediction. We are unaware of these correlations and their relative importance in comparison to other correlations for finding the answer we are looking for, yet these findings could all be subject to errors that might be ethically significant consequences. Normally, we would expect a developer to be responsible for inductive risk concerns such as these, yet due to black box opacity, they cannot manage these inductive risks.

---

53   Kearns and Roth 2020, 7–10.

This contributes to what Ryan Felder calls an "accountability gap" in cases where decision-subjects are concerned. An accountability gap exists when there is a vacuum created when a task is taken out of human hands and put into the hands of a machine.[54] The opacity of the model means that a developer or executor will be unaware of the statistical relationships that the model is looking for and hence unaware of whether there is good evidence for relying on them and their relative importance for producing a predicted output. Thus, not only is there a lack of accountability for someone like an operator who might be expected to explain the answer of a model to a decision-subject even though they don't understand the model, but there is a lack of accountability for the inductive risks that go into the model construction itself.

Let's consider a peculiar thought experiment. Imagine that we created an algorithm that could detect certain forms of cancer from a medical imaging device. The hope is that we could create a data set of image scans from this device and use machine learning to create a model that can determine which scans are cancerous and which ones are not. Ideally, our neural network will create a model that will generalize such that when we expose it to new scans it was not previously trained with, it will detect cancer accurately.

Now let's imagine that unbeknownst to us, the process involved in producing these medical image scans produces an artifact on the image that coincidentally happens to be present every time there is cancer and absent when there isn't cancer, despite the reason for the artifact's creation being unrelated to cancer. It might be so small and insignificant (like the size of a pixel) that we may not even realize it is there. The ANN producing the model picks up on this correlation and uses it to predict cancer; however, due to the opaque nature of the model, we are unaware of this.

So long as the model is exposed to scans produced by the same process, it would not only not be caught as a training error, but if the test data contains the same artifacts, it will not be caught as a test error either. If that unknown feature is present in the scan, the model will generate an accurate prediction. On this basis, the model is accepted and put into use in medical practices. Now imagine that we eventually change the process that produces those image scans and the artifact is no longer present. We wouldn't realize that there is a change, but without

---

54   Felder 2021, 40.

that key feature, the model will start producing errors and we wouldn't even know it because we trust the model but don't know it was looking for an irrelevant feature in the first place.

A thought experiment like this reveals the ethical risks involved when machine learning can pick up on correlational relationships that can seem to yield accurate results, yet in reality be meaningless. Moreover, the developer might not know about this and it wouldn't be revealed as training error or test error. If we accept the model and put it into widespread use and then later these features change, we may not realize that the model might start producing false negatives. The morally salient consequences of these errors might not be realized until a long pattern of errors is detected (which may not be believed if we trust the model). This makes it incredibly difficult to manage the ethical risks of error.

This thought experiment is a bit far-fetched since it assumes a purely coincidental artifact will affect the model construction in a very specific way. Is there evidence that real algorithms rely on useless information to seemingly provide accurate predictions in this way? According to Daphne Koller of Stanford University, such cases can happen. She explains:

> Imagine that you're trying to predict features from X-ray images in data from multiple hospitals. If you're not careful, the algorithm will learn to recognize which hospital generated the image. Some X-ray machines have different characteristics in the image they produce than other machines, and some hospitals have a much larger percentage of fractures than others. And so, you could actually learn to predict fractures pretty well on the data set that you were given simply by recognizing which hospital did the scan, without even actually looking at the bone. The algorithm is doing something that appears to be good but is actually doing it for the wrong reasons.[55]

Of course, the problem is that we may not realize that things like this are even happening owing to the opaque nature of the model.

Let's consider what happens when we can understand some of these models. In 2019, a team at the University of Tübingen created an image classification algorithm generated using a combination of a deep neural

---

55   Smith 2019.

network to recognize image parts combined with a transparent process that uses the number of detected features in an image to classify it. They trained this algorithm using images from a dataset known as ImageNet and then were able to have the algorithm point out what features in the image were most important for its decisions. Some of the results were surprising. For example, when it was asked what the most important features were for identifying a tench (a kind of fish), the result was that it was looking for human fingers against a green background.[56] All the images of a tench in the dataset included humans holding up the fish to a camera with their hands as a trophy, thus making it a predictive feature. An image-generating algorithm trained on the same dataset was asked to produce an image of a tench, and it produced photos of humans holding a fish with an emphasis on human fingers.[57]

In another example, a research team created an image classification algorithm that was trained to recognize different kinds of animals, including distinguishing between huskies and wolves. Using a process to try to detect which patch of pixels the model is using to classify images, they discovered that it distinguished between huskies and wolves not based on any features of the animal themselves but on whether there was snow in the background. If there was snow, the image was classified as a wolf, and without snow, it was classified as a husky (Figure 23). Why did this happen? All the images of wolves in the wild had snow in the background. As the team noted, "Often artifacts of data collection can induce undesirable correlations that the classifiers pick up during training."[58]

As Longino says, the universe does not come with labels. The way that a neural network might find patterns in the data to produce an answer means that they may find patterns that make little sense to us or would be difficult to explain. These models not only detect statistical relationships within the data, but they determine how significant those relationships are for producing an output. Yet, due to the opaque nature of the model we may not know which patterns the model is looking for or how important any of those patterns might be for producing a result.

This makes it very difficult to consider what counts as sufficient evidence for relying on these patterns. How important should the statistical connection be between one variable and another before you would use it

---

56   Brendel and Bethge 2019, 5.
57   Shane 2020.
58   Ribeiro, Singh, and Guestrin 2016, 1142.

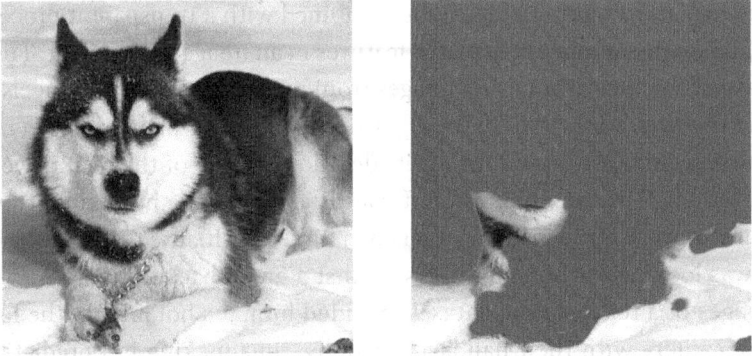

FIGURE 23 · Using techniques such as LIME, we can explain why a network classified an image a particular way. The model would have falsely concluded that this husky on the left is a wolf based on the snow in the background rather than any features of the dog. If we aren't careful with our training data, a neural network can rely on meaningless correlations to generate predictions that might appear accurate.

in the world? Also, input data itself is also subject to error. Credit reports (summarizing a person's record at payback of loans, etc.) are a commonly used piece of input data in machine learning, but studies show that more than a third of credit reports contain errors.[59] If the reliability of a given piece of data is in question and on top of that you don't know how important that particular variable is in terms of producing an answer, how can you manage inductive risks associated with the use of that data?

INQUIRER'S TOOLBOX

7. Am I considering all the information relevant to a solution? Is that information reliable?
8. When I consider how chosen ends might function as means to future ethical situations, are there major ethical concerns to consider?

If we can explain why a model produced a particular answer given specific inputs, it is **interpretable**.[60] In order to manage inductive risks in the face of opacity, we might seek to make AI-generated models more interpretable. Is there a way to try to peek inside the black box and figure out what the model is really doing? If we can gain some understanding of these models, how can this insight inform solutions for our ethical problems concerning the risks of error?

---

59   Gill 2021.
60   Ribeiro, Singh, and Guestrin 2016, 1136.

3.5 PEERING INSIDE THE BLACK BOX

There are several ways humans can gain understanding of how large and complex neural networks function. Arriving at a global understanding of the entire model is going to be difficult. However, we can attempt to understand how specific decisions or specific kinds of decisions were made. For example, we can create an algorithm that will provide specific kinds of variations in inputs to examine counterfactual considerations.[61] If small changes are made to specific variables, for example, does this greatly affect the outcome of the answer?

One way to understand which patterns neurons in the hidden layers are looking for is to try to learn the individual activations and weights for each neuron. If we know what the optimal stimuli are, we might get a good sense of which patterns the neuron is looking for. In 2012, Google created a large neural network involving over 16,000 CPU cores and trained a model with over one billion connections using data from YouTube to classify images. With such a large network, they were able to detect the optimal stimuli for a neuron and visualize it.[62] The result was a neuron that had clearly learned a human face or a neuron that had learned the face of a cat.

Nevertheless, not all networks can detect optimal stimuli in this way. It is rarely useful to cite the values of individual weights and connections. The complexity of these models often means that there is no way to predict results even with complete knowledge of the parameter values, and there is often no way of knowing in advance if a single parameter will change the relevant system's behaviour entirely or in a way that is imperceptible.[63]

Input heatmapping is another technique for highlighting features of an input to a network that are predictive of an output. This is particularly useful for networks that classify pixelated images according to the features of the image. Heatmaps emphasize the pixels or pixel regions that are most responsible for classification by tracing the responsibility that individual neurons have for producing activity in further layers in the network.[64] When these neuron activations correspond to pixels of

---

61   Kearns and Roth 2020, 174.
62   Le 2013, 8597.
63   Zednik 2021, 280.
64   Zednik 2021, 275.

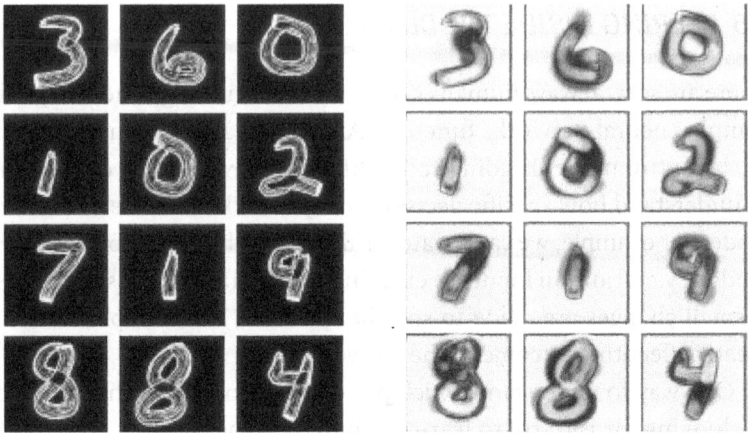

FIGURE 24 · Input heatmaps like the images on the right can be generated for various input images like the ones on the left using a process called layer-wise relevance propagation to indicate which areas of the image are more relevant and which are less relevant for producing an output.

an input image, a heatmap can be created to indicate those regions that are more relevant for determining a classification and which areas are weighed against that classification (Figure 24).

Techniques like these might allow us to verify that a network is looking for the right kinds of clues to generate classifications rather than spurious correlations. But there is no guarantee that even if the portions of an image that most strongly determine the network's output are highlighted, they will be clearly understood. The highlighted elements must look like some recognizable feature of the learning environment. However, since machine learning can track very subtle and complex features in the environment, they may not be easily recognized or are labelled by humans.[65] For example, a tiny unknown artifact like the one from our thought experiment still might not be revealed this way. Also, merely knowing where the network is looking within the image does not tell the user what it is doing with that part of the image.[66]

Another technique developed in 2016 aims to explain the predictions of any classification network by explaining the specific variables used to produce specific answers using qualitative textual or visual artifacts that indicate what produced the outcome.[67] Known as Local Interpretable

65   Zednik 2021, 277.
66   Rudin 2019, 210.
67   Ribeiro, Singh, and Guestrin 2016, 1135.

Model-Agnostic Explanation (LIME), this algorithm identifies an interpretable model that corresponds to how the model behaves within the vicinity of certain kinds of scenarios. In other words, while a large and complex model can be difficult to understand, especially for all possible inputs, instead we can construct an interpretable model that we can understand which we think approximates how the more complex model works within more limited ranges.

In one paper, researchers applied LIME to a text classification model trained to differentiate between "Christianity" and "Atheism" with a 94% accuracy rating. They discovered that the explanation for these differentiations seemed quite arbitrary, with words like "posting," "host," and "Re" having a significant effect on the outcome, despite these words not being clearly relevant to Christianity or Atheism. When they applied LIME to an image classifier, it was able to identify groups of pixels that were more strongly weighted towards producing an answer similar to Figure 23.[68] When applied to a model used to predict medical illnesses, LIME might be able to produce the specific symptoms in a patient's medical history that led to the diagnosis.

Tools like LIME could help us explain the results of an algorithm and thus to better manage our ethical responsibilities when it comes to our beliefs. By identifying irregularities that lead to erroneous outputs, it can "allow users to assess trust even when a prediction seems 'correct' but is made for the wrong reasons."[69] There are risks with using LIME, however. It will only capture a system's behaviour or a limited change of inputs, and so the approximations of the models it produces can risk being oversimplified.[70] There's also the fact that as an approximation, it may still not reveal all the ethically relevant information we might want from the model.

A 2020 paper even demonstrated a way to take advantage of techniques like LIME to effectively game the process and craft an arbitrarily desired explanation. It posits a malicious actor who intends to release a racially biased loan approval algorithm into the wild and hide this from a process like LIME. They could take advantage of the way in which LIME varies individual instances of input data to generate its approximation so that it would be very biased but would act very unbiased when dealing

68   Ribeiro, Singh, and Guestrin 2016, 1138–39.
69   Ribeiro, Singh, and Guestrin 2016, 1142.
70   Zednik 2021, 277.

with varied inputs from LIME. The result is that LIME would be fooled into thinking that a biased model is fair.[71] The paper's authors explain: "These results suggest that it is possible for malicious actors to ... effectively fool existing post hoc explanation techniques. This further establishes that existing post hoc explanation techniques are not sufficiently robust for ascertaining discriminatory behaviour of classifiers in sensitive applications."[72]

In terms of helping us resolve our ethical doubts about belief in the face of opaque AI, such techniques might be helpful, but they also add layers of inductive risk concerns in terms of how reliable they might be and whether they will provide all the relevant information we might require. An explainable model that is in 90% agreement with the original model explains the model for the most part but will still be wrong 10% of the time.[73] Of course such techniques to better understand these black box models continue to develop, but the ethical issue is not a mere theoretical one. The ethical question isn't whether a model might be theoretically explained one day; it is whether we can explain the models used today that produce significant ethical consequences for real people. How concerned should the developer be about what their model might do in the world?

3.6 ETHICAL STANDARDS IN THE FACE OF OPACITY

Given what we've learned about the nature of opaque AI, much more of the machine learning ecosystem is likely opaque to developers than we originally thought. Not only is the model a black box, but it has the potential to produce erroneous results in a way that might not be detected. This means that many of the potentially morally salient consequences of the model are also unknown. A model might rely on certain data points in unexpected ways and, unknown to the developer, it is possible that the input data itself can be subject to error, producing further errors still. Combined with the fact that some of the background assumptions might not be transparent as well and a developer might be far more unaware of what is going on than they might think (Figure 25).

---

71   Slack et al. 2020, 182.
72   Slack et al. 2020, 180–81.
73   Rudin 2019, 209.

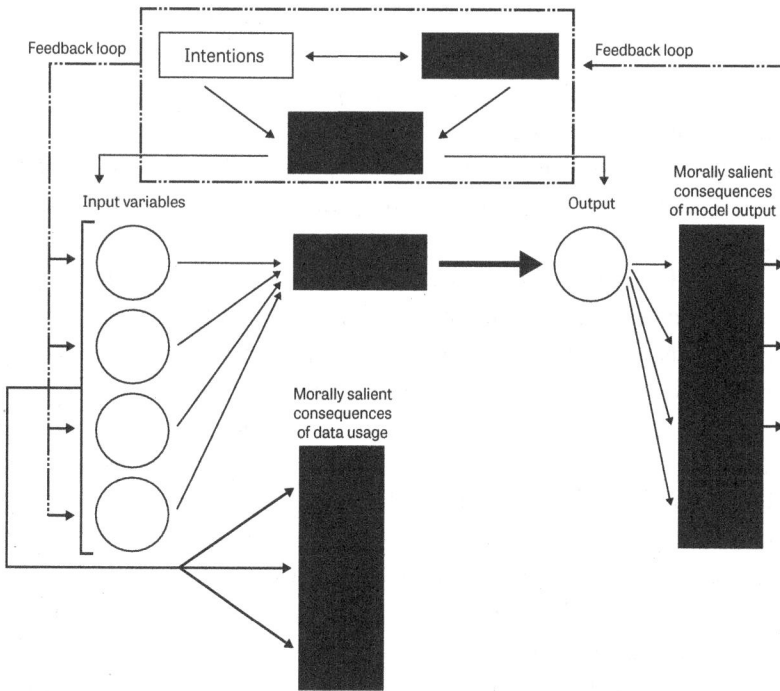

FIGURE 25 · In contrast to Figure 16, we know now that a creator or developer may be far less aware of what is going on within a machine learning ecosystem that we originally surmised. The issue goes far beyond the opacity of the model itself. Does this kind of opacity make it reckless for us to release the model into the wild?

All of this makes it very difficult to manage risk. As Professor of Computer Science Cynthia Rudin explains, "In high stakes decisions, there are often considerations outside the database that need to be combined with a risk calculation ... But if the model is a black box it is very difficult to manually calibrate how much this additional information should raise or lower the estimated risk."[74]

By contrast, we expect a scientist to avoid making reckless conclusions and they can do so by changing standards of statistical significance. Even then, the process can be faulty, as we see with the reproducibility crisis. An AI developer does not have the option of carefully controlling for every statistical correlation in this way because machine learning is automated. This highlights the central ethical issue: how much in the

---

74   Rudin 2019, 211.

dark does the developer have to be about what their models are doing before it becomes reckless or negligent to allow their use in the wild?

Using cost-sensitive machine learning can help build inductive risk concerns into model construction, but it doesn't alleviate all the issues that follow from opaque machine learning. This is also why Karaca highlights the difference between assessing a model during its construction and doing so in terms of accepting it before putting it in the wild.[75]

We might say these issues are simply an overfitting problem. In our thought experiment, the model would not rely on the scan artifact if varied kinds of input data were used. An algorithm wouldn't learn to detect wolves from huskies using the background if a more diverse set of photos was used during the training process. If more photos featured the tench in water during training, a model would be less likely to confuse fish with fingers. But part of the problem with opacity is that you never fully know what kinds of meaningless correlations a model might pick up on, so you can never fully anticipate in advance where overfitting on specific features might occur. In essence, this means that when a developer claims that their model isn't overfitting, they are simply making a bet that any new data fed into a model once it has been accepted won't be significantly different from the training data in ways they can't fully anticipate. Thus, there is still always a significant inductive risk concern.

Given this, one might wonder if using machine learning that produces opaque models should be acceptable, particularly for ethically sensitive issues. Perhaps we should only use models that are interpretable or, in other words, understandable to humans. As Rudin argues, in cases of high-stakes decisions, we should not use a black box model when an interpretable model exists with the same level of performance.[76] While there are times when we might prefer an opaque model to an interpretable one, such as in cases where we are concerned about the security of the model, Rudin argues that we are better off using interpretable models such as decision trees or linear regressions to make decisions. Here, however, there is a trade-off since building interpretable models requires greater resources and expert understanding of whatever is being modeled.

---

75   Karaca 2021, 6.
76   Rudin 2019, 211.

Depending on where you happen to be in the machine learning ecosystem, you might demand a different standard of interpretability according to what is epistemically relevant to you. As Zednik points out, an end-user might be more preoccupied with "what" questions or "why" questions; what is the model doing and how does it translate these inputs into a particular output? Or why is this an appropriate model for this particular environment? On the other hand, a creator or developer might be more interested in a "how" question, or a question pertaining to how to intervene in the system to get the network to change its behaviour.[77] Depending on what you are interested in there may be many ways to render a model interpretable.

Is there ethical justification for using these models, particularly if we cannot manage inductive risk concerns? Without an explanation of how the model works, all we can use to evaluate the model is to look at its predictive accuracy. How well does the model perform on test data? How many mistakes does it make? But if you are unaware of the full extent of what is going on or what mistakes could be made, is it ethical to believe it? After all, we already know that algorithms can seemingly generate answers that appear consistently correct yet are based on meaningless statistical correlations. Or, when we use opaque models to help make difficult decisions despite not knowing if they work, do we risk, as Clifford suggests, believing for unworthy reasons, and do we risk undermining our own powers of self-control and for judiciously weighing evidence before we reach a conclusion? Remember that a bad algorithm can seemingly generate accurate results just as a dilapidated ship might make it to port.

4. Opacity from the Perspective of Decision-Subjects

To consider what our ethical responsibilities are when it comes to believing and acting on the conclusions of AI as decision-subjects or end-users, we'll examine a few case studies where an algorithm might be used, what kinds of conclusions it might reach, and whether we as end-users should trust and act on the conclusions of AI based on what we might not know.

---

77   Zednik 2021, 278.

4.1 HIRING AND EMPLOYMENT

Companies that are hiring can be overwhelmed by the sheer volume of online applications. Many companies use algorithms and applicant tracking systems (ATS) to automatically review applications and suggest potential hires. Three quarters of all résumés are no longer read by human eyes.[78] In Chapter 3 we discussed how such algorithms can be problematically biased. But they can also be ethically problematic because of the opaque assumptions and correlations that the algorithm might rely on. You may be evaluated using metrics in ways that are not transparent to you, relying on correlations nobody is aware of and leaving you little recourse for appeal.

Imagine that you want to create an algorithm to help you find quality job hires. What metrics would you look for and what background assumptions would justify them? Would you consider how long they've worked at previous jobs? Would you consider where they live? How would you estimate how reliable they might be as an employee? Would you use a credit score? As we've discussed, training an algorithm requires a lot of data and so you'd want to find metrics that you think might indicate a good hire at scale. Many corporations now make use of psychometric assessments in the form of different personality or aptitude tests. Such tests can be used to determine your cognitive ability, personality and temperament, or even potentially for medical diagnoses. Use of such a test at scale by all applicants would allow an algorithm to quickly screen them.

Many psychometric tests, such as the Raven's Progressive Matrices and the NEO Personality Inventory, have been used in the social sciences with relative success. In hiring applications, the typical personality tests include the Myers-Briggs Type Indicator, the DiSC model, and the Big Five model.[79] These tests seek to measure everything from how open to new experiences you are, whether you are extraverted, or agreeable, or how you respond to new problems. Corporations selling AI hiring solutions promote their own proprietary tests that rely on metrics that they believe can signify a quality job hire. These might include looking at social media postings, linguistic analyses of writing samples looking

---

78   Renzulli 2019.
79   Meinert 2015.

for certain keywords associated with success, or video-based interviews that use algorithms to analyze speech content, tone of voice, emotional states or temperamental cues.[80]

There are several potential problems with the use of such tests. To begin with, you as an applicant may not know what the test is testing for, or its overall weight in evaluation. If the test is inherently unfair, for example, you might be rejected because of this but never know about it. Would it be ethical for a company to assess your tone of voice from a recording by asking you to participate in the recording but not tell you why? In 2020, a makeup artist was furloughed during the pandemic and told to re-apply for her job. She tested well, but the AI tool scored her body language poorly and she lost her job. She noted that job candidates rarely know if these tools are the reason companies reject them and they aren't allowed to see how they've been evaluated.[81] In another case, someone who had been screened out from one application reapplied, changing their birthdate: they got an interview.

Kyle Behm sued a company named Kronos after he discovered from a friend working at the company that he had not been hired because of his personality test. Being bipolar, he recognized some of the questions from personality tests he had taken in previous mental health assessments.[82] Despite having good grades and going to a good school, Kyle could not get hired.[83] It is illegal under the Americans with Disabilities Act to use a medical exam for hiring purposes. Disabilities, whether mental or physical, are considered private information that an employer cannot ask about at the pre-employment stage. So, the issue is what these tests are for and how they are being used.

The opaque process could be using proxies to discriminate against groups of people, (intentionally or not). Some personality questions are proxies for personality types, but they can also signify mental illnesses and other issues. Questions about emotional states and mood swings, for example, might signify mental illness and be used as a basis of weeding out certain applicants. A woman named Vicky Sandy sued a supermarket after taking a Kronos personality test. The test evaluated her for factors like patience, being a team player, and attentive and respectful listening.

---

80   Dattner et al. 2019.
81   Lytton 2024.
82   O'Neil 2018.
83   O'Neil 2016, 106.

Sandy is hearing and speech impaired and scored low on the test and, as a result, it suggested the interviewer listen for "correct language" and "clear enunciation."[84]

Another problem is that the background assumptions justifying the use of these tests in this way and for thinking that such tests signify anything in the real world may be bogus. As a *Harvard Business Review* article put it, "Many of these tools have emerged as technological innovations, rather than from scientifically derived methods or research programs. As a result, it is not always clear what they assess, whether their underlying hypothesis is valid, or why they may be expected to predict a job candidate's performance."[85] Why, for example, should we think that there might be a connection between your tone of voice and having a certain kind of personality, and that moreover this will predict future performance? Psychological studies reveal that relying on personality tests alone isn't very reliable for making effective hiring decisions.[86] If the test isn't public, the basic assumptions may go unchallenged and you might have to submit to an opaque pseudo-scientific process.

Neither does the use of algorithms stop once you are hired, for as we noted at the beginning of the chapter, such algorithms (including their potentially problematic background assumptions, limited or biased data, and lack of transparency) can also be used to evaluate your employment performance.

4.2 E-SCORES AND LOANS

In the field of e-commerce, deciding how much credit someone is worth, or whether someone should get insurance, can be a risky proposition. If you lend the money and the recipient is unreliable, you lose out. However, an algorithm will allow you to judge applicants much more efficiently.

Traditional lenders like banks only use a borrower's credit score to determine creditworthiness. Credit ratings are transparent, in that the understanding of how to use them and improve them are well known, and you have the right to ask for all the information that goes into your credit score. However, a company may wish to use their own risk evaluation that they believe is more comprehensive, or they may be legally

---

84   Ellin 2012.
85   Dattner et al. 2019.
86   Prospect AI 2019.

prevented from using credit scores for marketing purposes. Thus, as discussed in Chapter 3, they will instead use different proxies and compile them into a risk assessment called an e-score.[87] This might include such variables as where you live, your employment history, purchase history, internet surfing patterns, and even social media presence. For example, your score could be lower if you engaged in "condemnable activities like stalking or dropping inappropriate comments."[88]

These scores involve taking different pieces of personal information and correlating them with measures of success and reliability by essentially comparing you to people who are statistically like you. So, in essence, the score is no longer about what you do and who you are, but about what the group you are part of does. Worse still, e-scores are invisible: you will probably never know your e-score.[89] As Cathy O'Neil, explains, the result is that "e-scores are arbitrary, unaccountable, unregulated, and often unfair."[90] Further, because of the opaque nature of the score, you have little recourse to complain if it produces a poor evaluation of you.

Your e-scores may also be used when you apply for a personal loan, a credit card, or car insurance.[91] This also means that information may be held against you in ways that are arguably irrelevant. They are based on information about you and people like you and will be compiled to evaluate you in ways that you are unaware of, and that are often unregulated and unethical, ways that may rely on pseudo-scientific ideas about the meaning of various data-points and their relation to financial reliability.

4.3 MEDICAL DIAGNOSES

Medical services are incredibly expensive and there are often limits on resources. AI can be applied in the medical field to make healthcare delivery more efficient. Some algorithms, as we've discussed, are responsible for determining the level of resources allocated to specific patients, while others can be responsible for looking at medical scans and making diagnoses.

---

87 O'Neil 2016, 145–46.
88 Singh n.d.
89 Singer 2012.
90 O'Neil 2016, 143.
91 Evalueserve 2022.

AI has the potential to make healthcare delivery more efficient, given chronic staff shortages and a lack of resources. However, safely implementing AI technologies in the medical field may be tricky, given the black box problem and algorithmic opacity, as we know that we sometimes cannot determine how these technologies arrive at their results and hence the question of justification remains.

In some cases, because these algorithms are trained on massive data sets, we cannot be sure they are picking up on what is relevant to make a correct or accurate diagnosis. For example, an algorithm designed to screen for malignant skin cancers was found to correlate cancer with the image of a ruler in a photo.[92] Why rulers? Physicians and clinicians routinely measure the size of tumors using rulers to create a visual comparison, and the technologies were trained on data sets with images of rulers. So, the algorithm honed in on that pattern: rulers and skin cancer go together. Of course, we know that rulers don't cause cancer or have anything to do with its incidence. The pattern was accidental.

Should we trust results if we do not know how something works? Should we trust them even if the technology occasionally makes mistakes, such as associating rulers with cancer? As Ryan Felder argues, not knowing how something works or exactly why it is effective doesn't necessarily mean we lack the justification for using it.[93] Consider lithium, effective in treating individuals with bipolar disorder and other mental illnesses. It works, but we have yet to figure out precisely how it works. However, to deprive individuals who benefit from it because physicians or scientists lack a thorough understanding of how and why it works would be harmful to those who rely on it.

Similarly, human radiologists often can't explain how they have arrived at a diagnosis. They can point to problematic areas of x-rays, but they usually have no theory explaining why a region is problematic. Their ability to see involves perceptual sensitivity, but they cannot fully explain themselves. This raises an important question about whether we might unfairly

INQUIRER'S TOOLBOX

Are there analogies that can inform how I might understand a problem or a potential solution? In what ways is that precedent helpful or relevant? In what ways is it not?

---

92   VentureBeat 2021.
93   Felder 2021, 39.

use a double standard against AI by considering their opacity to be ethically problematic while overlooking human-derived opacity. Is there an
ethical difference between the two?

When we ask ourselves if we know something, there is a traditional
assumption that we must be able to account for why we know it.[94] We
must be able to access and reflect on our reasons; otherwise, we do not
have justification for what we claim to know. However, some philosophers argue that we are justified if we use reliable mechanisms appro-
priately linked to that truth, whether one is aware of them or not. We
can still gain knowledge even if we are unaware of our reasons and have
mechanisms that get things right cognitively. For example, our visual
perceptions are generally a reliable process for forming beliefs, even if
we cannot articulate why our vision is reliable. But our beliefs formed on
the basis of visual perception are justified because our vision is reliably
conducive to truth. And so, according to this view, we can be justified on
that basis. The point here is that even if we fail to know the inner workings of something or lack awareness of the reasons, this does not rule
out justification (of the sort that relates to knowledge).

We might ask whether the statistical validation of the black box technologies is a sufficient justification to use them. Felder maintains that
we cannot have vanishing accountability in using AI algorithms and
automation in healthcare.[95] Someone must take responsibility for their
implementation.

To meet the condition of accountability, one must ensure the systems
are operating as they should and that someone is responsible if things
go wrong. Accountability and retaining trust are cornerstones of healthcare. One proposal might be to certify such technologies. We often have
certification programs that require checks and measures when safety is
an issue. However, given algorithmic opacity, this is limited. Arguably,
a track record of success matters, even in the testing phases. Still, algorithms remain somewhat unpredictable and may forge false associations. Accordingly, companies will likely hesitate to certify these systems
and technologies, and the worth of such certification may be limited.

If we can't explain what justifies the conclusions of an algorithm,
we are left with the fact that statistically these algorithms can identify

---

94   Pappas 2014.
95   Felder 2021, 42.

correct diagnoses faster and sometimes more accurately than humans. Statistically, we might be better off with the algorithm, yet it will never be able to explain itself or be accountable for errors. And this raises some important questions. Is the fact that an opaque algorithm may become statistically validated by virtue of its ability to statistically perform better than a human a sufficient justification for our beliefs, or should we demand something more? If there is no way to hold anyone accountable for the algorithm's errors, is it still ethical to use it?

4.4 THE ETHICAL MEANING OF OPACITY FOR THE DECISION-SUBJECT

We've considered some of the ethical concerns relating to developers, but also what ethical challenges opaque algorithms offer the decision-subject or end-user who must face the consequences of an algorithm's conclusion.

When dealing with AI that can affect one's life, we must accept to a significant degree that for the end-user, the algorithm is largely a complete black box. In some ways the black box issue is significant for the developer, but it is also important for the end-user. You will be asked to accept the conclusions of an algorithm (whether that includes believing a diagnosis or accepting a YouTube recommendation) when you lack clear reasons for that conclusion. Operators and executors in a machine learning ecosystem may not be able to explain to you why you are being evaluated the way that you are. You may be unaware of what sorts of correlations a model may be looking for and these correlations can sometimes make no sense. The model may secretly be very biased and the corporation that owns it might keep that from you. This raises questions about your ethical obligations not to believe things when you have reasonable doubts.

Unless the institution who created the algorithm is extremely transparent, end-users are not likely to know the background assumptions involved. Even if you knew, for example, all the variables that go into the model or the outputs that the model produces, you might still not understand their full significance without the background assumptions. Remember that background assumptions justify why we take evidence of one thing as indicating another. You could know that a corporation is asking for a credit score, but not understand that it is using that as a proxy for your reliability as an employee. Those assumptions may also make no sense or be unfair. As O'Neil explains, "embedded within these models are a host of assumptions, some of them prejudicial."[96] Further, the developers themselves may not be fully aware they are relying on them.

You might not understand all the data that is being used in the model. You might not be aware that your internet surfing habits could affect your ability to get a job or a personal loan. You may not be fully aware of where the data comes from that is training the algorithm and whether it was reliable, was collected ethically, or whether the choice of input variables make sense given the background assumptions and the intentions of the developer. Models "place us into groups that we cannot see, whose behaviour appears to resemble our own."[97] You also will not likely know, unless the organization is transparent or there is some kind of investigation, whether the data being used contains biases. You also might not fully understand the intentions of the person who created the model. A hiring algorithm seemingly designed to find the best candidates might be simply trying to exclude as many people from the application process as cheaply as possible.

The fact that we understand so little about these systems yet are expected to accept the conclusions of an algorithm that could be in error is ethically problematic. And yet this problem is magnified by the fact that often the use of the algorithm introduces a double standard. As O'Neil points out:

> An algorithm processes a slew of statistics and comes up with a probability that a certain person might be a bad hire, a risky borrower, a terrorist, or a miserable teacher. That probability is distilled into a score, which can turn someone's life upside down. And yet when the person fights back ... the case must be ironclad. The human victims ... are held to a far higher standard of evidence than the algorithms themselves.[98]

This fact highlights once again the ethical responsibilities of end-users when it comes to what they are willing to believe and accept without additional explanation.

---

96 O'Neil 2016, 25.
97 O'Neil 2016, 164.
98 O'Neil 2016, 10.

AI always has the potential to make an erroneous judgement. Despite the opaque nature of models that much AI is built around, we might accept the judgement of an AI despite not understanding it if the algorithm has a proven track record for accuracy. But is statistical validation of an AI that consistently gets the right answer enough? If the models that enable self-driving cars were completely inexplicable to you, yet such vehicles were statistically less likely to get into an accident than a human driver, would that be sufficient to justify having them on the roads? Let's assume that future self-driving cars do perform better overall than humans, yet occasionally do cause accidents. Is it better to have black-box drivers who perform statistically better than human drivers who can explain themselves and be held accountable?

As Clifford states when it comes to the ethics of belief, "The question of right and wrong has to do with the origin of his belief, not the matter of it; not what it was, but how he got it; not whether it turned out to be true or false, but whether he had to believe on such evidence as was before him."[99] As end-users we run the ethical risk of letting our guard down in the face of sloppy statistics and engineering and accepting beliefs and pseudo-science. The more people accept opaque systems and the double standards involved when trying to appeal erroneous results, the more difficult it will be to challenge the conclusions of an algorithm. This is why Clifford adds, "Every time we let ourselves believe for unworthy reasons, we weaken our powers of self-control, of doubting, of judicially and fairly weighing evidence."[100]

While there are cases where if something is done by human hands or eyes, a complete explanation cannot be given either, those cases can still allow us to assign responsibility for error in ways that we may not be able to do with AI. Furthermore, one of the most significant differences between human cases of opacity and AI cases of opacity is that AI will be capable of operating on a scale that humans are not (that's why they are considered more efficient, after all). So, the essence of opacity is not just a lack of accountability but the potential to magnify errors in ways that can be difficult to detect and harder still to appeal or hold anyone accountable for.

---

99   Clifford 1879, 178.
100 Clifford 1879, 185.

5. Conclusion

AI has the potential to make our lives better. Not only can algorithms allow us to evaluate complex questions in efficient ways, but they offer the possibility of obtaining more accurate results. However, efficiency and ease come at the cost of opacity and a loss of ethical accountability. Many machine learning models are black boxes, making associations and connections in ways that may never be fully understood. But within a machine learning ecosystem, opacity can extend much further than just the black box that is the model. A developer may not be aware of a great deal of information that makes their own algorithms work. This can include important information that might shield a developer from understanding why their algorithm might make systematic errors. If a developer is ethically responsible for the errors of their work, then opacity within the system will make it difficult for them to manage inductive risk.

If you are a decision-subject, the machine learning ecosystem will likely be mostly opaque to you. You may have to submit to a poorly thought-out process utilizing a model based on faulty background assumptions and pseudo-science and you'd never know it. An algorithm could be generating errors (even if they seem like the correct answers) and you and the developers may not know about it. A corporation could even intentionally build a biased algorithm and hide this fact from you and from regulators.

If a decision-subject, an operator, or an executor accept the results of an algorithm, they are adopting a belief for which there may be no clear explanation other than the fact that the algorithm has been successful in the past. If statistical validation is necessary but not sufficient, what uses of AI are ethically permissible if we can't establish a stronger basis for an explanation? Should application matter? Can we accept opacity in the case of hiring algorithms but not the self-driving car? It's unlikely that a single ethical standard would apply equally in all use cases because the consequences of error can vary. In some cases, we might also have a higher expectation of accountability, and that might mean that there are some use cases where the opacity of the model would preclude its ethical use.

ADDITIONAL MATERIAL

Opacity • a lack of transparency; in artificial intelligence this includes the inability to explain how an algorithm works or why it reaches a conclusion.

Artificial neural network • a type of machine learning model that mimics brain function by developing a set of layered and interconnected neurons to solve problems.

Deep learning algorithm • a type of machine learning model that uses artificial neural networks containing hidden layers.

Black box • exists when we understand the inputs and the outputs of a model but don't understand how inputs produce the outputs.

Machine learning ecosystem • the system of actors interacting with a machine learning model in various capacities, including its creators and developers, operators, executors, decision-subjects (or end-users), data-subjects, and regulators.

Evidentialism • the principle that one is justified in believing something only if they have sufficient evidence to support that belief.

Null hypothesis • the claim that there is no statistically significant relationship between two processes or properties.

Statistically significant result • a result that is unlikely to be a coincidence based on assumptions about the underlying distributions of events.

Training data • labelled observations with known outcomes used to develop a machine learning model.

Test data • a subset of the labelled data with known outcomes that the model was not originally trained on. Used to test how well the model generalizes to new data.

Cost function • a calculation of a model's overall error, where error is the difference between the labelled outcome and the algorithm's predicted outcome.

Underfitting • occurs when a model is too simple and unable to capture the complexity of the statistical relationships in the training data.

Overfitting • occurs when the model conforms too closely with the statistical relationships in the training data and is unable to generalize to other test data.

Gradient descent • an optimization algorithm that seeks to minimize the cost function.

Cost-insensitive machine learning • the optimization of an algorithm applies equal costs to different types of training errors (e.g., false positives and false negatives) to maximize overall accuracy.

Cost-sensitive machine learning • the optimization of an algorithm applies different costs to different types of training errors (e.g., false positives and false negatives) depending on which error we wish to avoid most.

Interpretable • when a human can understand the reasoning behind the predictions that the model makes.

---

1. What is the black box problem and what does opacity mean in the context of AI ethics? Explain this concept in your own words and provide an example. What are some of the problems that algorithmic opacity raises?

2. Consider again W.K. Clifford's form of evidentialism. Why is it wrong to believe against or without evidence? How much evidence is sufficient to give one the right to believe what an AI tells them?

3. Some argue that we should implement algorithms in healthcare and medical settings, even if how they work remains opaque. However, even supporters of using such technology in healthcare may raise concerns about applying it to other areas, such as law and the legal system. Is the opacity issue more problematic in some fields or areas than others?

4. If you had no idea how a process worked, but nevertheless this process had a good statistical track record of success, how confident would you be in it? Would the risk of error in some circumstances make you demand a higher standard of evidence?

5. Are there some applications for AI that should be banned owing to opacity issues?

6. If malevolent actors could shield biases from processes that seek to explain what an opaque model is doing, how should regulators and end-users respond to this?

7. Is there a risk that the efficiency of AI will make us overlook the odds that it will generate error and that we will become too credulous for AI-generated answers?

5

Democracy

The world is rapidly changing. New inventions and technology threaten to change the world in profound and unforeseeable ways. Rapidly changing economics and the limits of resource extraction have created economic and political instability. The prospect of war using new technologies capable of causing untold horrors is becoming an increasing possibility. Novel political movements considered to be at the extreme ends of the political spectrum have taken root and are growing in popularity. This is partially owing to new technology, which has enabled mass communication and the spread of misinformation and propaganda. Such conditions have raised concerns about the stability of the democratic world and whether it can respond to and manage such crises.

You might think that I am speaking of today, but I am speaking about the world of the 1920s. After World War 1, the rise of assembly-line production, mass communication, and new technology profoundly changed the world. The automobile was being popularized and air travel was undergoing rapid development. Film and radio allowed communication on a never-before-seen scale and revolutionized advertising, giving rise to a mass consumer culture and the use of political propaganda. The world was also becoming a more technical and scientific place and to many it seemed that the world was too complicated, and that what was needed was for experts to direct policies. As the 1920s gave way to the depression of the 1930s, growing political and economic instability led to calls for reform.

Given these concerns, some questioned whether democracy was still well-suited to a changing world. Some questioned the nature and function of democracy and whether some form of expert rule might be necessary. In his 1925 book *The Phantom Public*, Walter Lippmann explained that

> When public opinion attempts to govern directly it is either a failure or a tyranny. It is not able to master the problem intellectually, nor deal with it except by wholesale impact. The theory of democracy has not recognized this truth because it has identified the functioning of government with the will of the people. This is a fiction.[1]

Today mass communication is even more 'mass', with information (genuine or otherwise) instantly able to travel around the world and be read by millions. We have greater control over exactly what news and information we wish to see, even if it only reinforces our biases. AI promises to bring exponential change with political and ethical implications that require technical expertise that the public may not be capable of fully understanding. If we do have to rely on experts to govern AI use, does this challenge the principle of self-rule? If we want to determine if AI is a threat to democracy, however, we will first need to better understand democracy.

These issues bring to light some central questions that this chapter will explore:

1. What is the nature of democracy?
2. Does artificial intelligence undermine democracy?
3. Does artificial intelligence undermine the public's ability to grasp reality and respond to common problems?
4. Does the highly technical nature of artificial intelligence and its relationship to long-term and wide-scale ethical challenges mean that we need to rely on expert rule to solve problems?
5. Is democracy still an adequate form of government if it proves unable to address modern challenges?

---

1   Lippmann 1993, 61.

6. Should political campaigns use artificial intelligence to help their chances of winning?

7. How should democracy and the political machinery that runs it be reformed to respond to new challenges posed by artificial intelligence?

The similarities between the past and the present might suggest that these problems are nothing new and that we needn't worry, but it is more helpful to consider the fact that these systemic problems facing democracy over the previous century have never been overcome, including determining how democracy should govern the uses of science. To that end, we will consider in further detail the cynical account of democracy presented by Lippmann and the public debate it sparked with John Dewey on the nature of democracy, and determine whether their views can offer insight towards resolving issues facing contemporary democratic societies in the face of technological development.

1. Democracy and the Public

What is democracy? If you look up a definition, you will find that it means rule by the people. It is the idea that political legitimacy stems from the consent of those who are governed and that the people should have a say in what laws are passed and who governs them. But what does this entail? Is the mere act of voting based on "one person, one vote" sufficient to say that a society counts as democratic, or is there more to it? Some might argue that democracy is impossible without the rule of law. Others might say that democracy is impossible without respect for human rights and civil liberties such as freedom of speech or freedom of conscience. Still others might even claim that democracy requires a commitment to certain kinds of social equality.

Democracy can be construed in a very thin sense or in a very robust sense. A key question for us to consider is whether democracy has anything to do with ethics. To determine if it does, we will examine the ideas of Jane Addams (1860–1935), a social reformer, sociologist, and philosopher, and co-founder in 1889 of Hull House, a Chicago settlement house designed to bring people of different classes together in close proximity.

Addams believed in cross-class contact and that antagonism, caused by the injection of personal attitudes rather than by objective differences, was always unnecessary.[2] When we gain a better understanding of each other's experiences and leave our personal feelings aside, we can cooperate and resolve our differences by arbitration. This was a part of a concept she termed **sympathetic understanding**, whereby we learn from each other's perspectives and experiences to better understand common connections and enable empathetic moral responses.

For Addams, ethics requires democracy and is attained by "mixing on a ... common road where all must turn out for one another, and at least see the size of one another's burdens." She adds, "To follow the path of social morality results perforce in the temper if not the practice of the democratic spirit, for it implies that diversified human experience and resultant sympathy which are the foundation and guarantee of democracy."[3] In other words, ethics requires a spirit of democracy, and both require that we adopt an attitude of sympathetic understanding to recognize and resolve each other's problems.

Addams's views on democracy influenced many scholars, including John Dewey, who explains,

> A democracy is more than a form of government; it is primarily a mode of associated living, of conjoint communicated experience. The extension in space of the number of individuals who participate in an interest so that each has to refer his own action to that of others, and to consider the action of others to give point and direction to his own, is equivalent to the breaking down of those barriers of class, race, and national territory which kept men from perceiving the full import of their activity.[4]

Thus, as Addams and others understand the concept, democracy requires the exchange of perspectives and experiences for the purposes of remedying social ills and to allow the individual to make a meaningful contribution to the greater whole of society.

Given this more robust conception of democracy, should we worry about AI threatening democracy and democratic norms? AI has a

---

2   Menand 2001, 313.
3   Addams 1902, 6–7.
4   Dewey 1916, 101.

worrying capacity to spread misinformation by generating fake but very realistic-looking content with speed and efficiency. It allows for the use of microtargeting, whereby political messaging can be tailored to specific individuals; this can contribute to the creation of filter bubbles.[5] A "filter bubble" is a state of intellectual isolation that shields people from information and argument with which they disagree. Can microtargeted advertising and filter bubbles undermine our ability to democratically engage with one another and for the public to manage its own affairs?

Addams would likely think so. She argues, "We know instinctively that if we grow contemptuous of our fellows, and consciously limit our intercourse to certain kinds of people whom we have previously decided to respect, we not only tremendously circumscribe our range of life, but limit the scope of our ethics."[6] Others worry that AI will be very difficult to regulate in practice, especially owing to a lack of understanding of the technology by the public and by lawmakers.[7] This means that the public will have to rely on experts to articulate specific laws and policies, something that can again undermine the democratic process.

If you believe in a very robust notion of democracy, then AI might seem like a significant threat. If the public can't understand the issues because of misinformation or because filter bubbles prevent any kind of common experience, or even because the issues seem too complex and remote from everyday life, then this will undermine democratic efforts. But perhaps we are romanticizing democracy if we place too much emphasis on the public's role in governing its own affairs. To consider a different perspective, we will return to Lippmann's views and consider whether his account is relevant to contemporary issues of AI and democracy.

1.1 LIPPMANN'S CYNICAL ACCOUNT OF THE PHANTOM PUBLIC

Walter Lippmann was a journalist who served as an advisor to the United States president on the issue of censorship during World War I. His 1925 book *The Phantom Public* outlines why he believes the modern

---

5   Heath 2023.
6   Addams 1902, 10.
7   Kang and Satariano 2023.

world has become too complicated and why the public is incapable of governing itself.

Lippmann argues there is nothing ethically significant about elections. He calls elections "sublimated warfare" held to avoid actual civil war.[8] Supporters of democracy falsely believe that democracy is about a public deciding for itself how it wants to be governed according to "mystical" notions that the public is engaged, interested, and capable of governing itself. Instead, Lippmann holds that "this public is a mere phantom. It is an abstraction."[9]

These notions of democracy, according to Lippmann, rely on an "omnicompetent, sovereign citizen" who collectively makes up the public, yet he argues this is an unattainable ideal.[10] One reason Lippmann believes this is that most people are indifferent to most public affairs, which are too complex and numerous to follow; or people simply don't have the time to be informed and engaged. The average voter cannot grasp the problems of the day and has no time to learn about them and no interest or knowledge of such things.[11] The amount of facts exceed our curiosity and the details of most public affairs are remote and too intricate for the average reader; hence "the rest of us ignore them for the good and sufficient reason that we have other things to do."[12] Lippmann notes how the arrival of mass communication increases the potential to be aware of public affairs, yet this does not happen in practice.

Lippmann was aware of how confusing and distracting mass communication could be:

> If by some development of the radio every man could see and hear all that was happening everywhere, if publicity, in other words, became absolute, how much time would the public spend learning about every issue? It is bad enough today—with morning newspapers published in the evening and evening newspapers published in the morning, with October magazines in September, with the movies and the radio—to be condemned to live under a barrage of eclectic information, to have one's

---

8   Lippmann 1993, 49–50.
9   Lippmann 1993, 67.
10   Lippmann 1993, 28–29.
11   Lippmann 1993, 27.
12   Lippmann 1993, 33.

mind made the receptacle for a hullabaloo of speeches, arguments and unrelated episodes.[13]

Given Lippmann's concerns about the potential for the public to keep
track of affairs in the age of mass communication in the 1920s and 1930s,
what would he think today?

Asking people to vote more often or be more engaged with something they don't understand won't help. "If the voter cannot grasp the
details of problems of the day because he has not the time, the interest
or the knowledge, he will not have a better public opinion because he is
asked to express his opinion more often," Lippmann argues.[14] To do so
would be a "compounding of individual ignorances in masses of people."
Public affairs are for the most part invisible, being managed behind the
scenes by vested interests.

If we reject "mystical" democratic accounts of how government
works for being too unrealistic, then how does Lippmann think government works? He explains:

> Actual governing is made up of a multitude of arrangements
> on specific questions by particular individuals. These rarely
> become visible to the private citizen.... The mass of people see
> these settlements, judge them, and affect them only now and
> then. They are altogether too numerous, too complicated, too
> obscure in their effects to become the subject of any continuing
> exercise of public opinion.[15]

We make no attempt to understand or consider society as a whole. Most
choices that the public makes are limited in scope, since "before a mass
of general opinions can even eventuate in executive action, the choice is
narrowed down to a few alternatives. The victorious alternative is exe-
cuted not by the mass but by individuals in control of its energy."[16]

Any role for public input will only follow once a range of opinions
have been simplified, unified, and formulated for the public to consider.
As Lippmann explains, the public "discerns only gross distinctions, is

---

13   Lippmann 1993, 34.
14   Lippmann 1993, 27.
15   Lippmann 1993, 31.
16   Lippmann 1993, 38.

slow to be aroused and quickly diverted; that, since it acts by aligning itself, it personalizes whatever it considers, and is interested only when events have been melodramatized as a conflict."[17] The public does not express opinions, but rather merely aligns for or against a proposal. Thus, "we must abandon the notion that democratic government can be the direct expression of the will of the people ... by their occasional mobilizations as a majority, people support or oppose the individuals who actually govern. We must say that the popular will does not direct continuously but that it intervenes occasionally."[18]

Clearly Lippmann is not an optimist about what democracies can aspire to. The public is largely too disengaged and ill equipped to govern itself. In his 1922 book *Public Opinion*, Lippmann argued that experts had to effectively govern for us. He explains: "common interests ... can be managed only by a specialized class whose personal interests reach beyond the locality."[19] However, in his later book he accepts that experts are fundamentally outsiders to the problem and thus can only offer indirect assistance.[20] While these people must govern, they are subject to the same faults as the rest of us, and so the function of the public is to reign in those who run society from arbitrary behaviour or lawless power.[21] As he explains, public opinion should "check the use of force in a crisis, so that men, driven to make terms, may live and let live."[22]

If we accept Lippmann's account, then many of the issues involving AI and democracy take on a new lens. If we begin to consider only some of the issues we've discussed in this book, is it reasonable to expect that the public will be engaged in regulating AI and will understand the issues? If the public is largely a phantom in any case, can AI really threaten the process of democracy?

1.2 DEWEY'S REBUTTAL: THE PUBLIC AND ITS PROBLEMS

Two years after Lippmann published *The Phantom Public*, John Dewey published *The Public and its Problems* in response. According to Dewey,

---

17 Lippmann 1993, 55.
18 Lippmann 1993, 51–52.
19 Lippmann 1922, 310.
20 Lippmann 1993, 62.
21 Lippmann 1993, 134–35.
22 Lippmann 1993, 61.

the public isn't a phantom but rather it "is so confused and eclipsed that it cannot even use the organs through which it is supposed to mediate political action and policy."[23] Dewey starts by recognizing that human acts have consequences. Once we understand an event as a consequence produced by certain actions, we can try to mitigate and control those consequences. When a consequence affects the person engaged in the activity, it can be called a direct consequence, whereas indirect consequences affect people beyond those who are directly involved. It is this distinction that gives rise to public activity.[24]

An action that only affects the persons directly engaged in it is private. When those consequences begin to affect the welfare of others, they acquire a public significance. The public consists of all those who are affected by indirect consequences to the degree that they take steps to control those consequences. As Dewey explains, "The characteristic of the public as a state springs from the fact that all modes of associated behavior may have the extensive and enduring consequences which involve others beyond those directly engaged in them."[25] A public forms once the recognition of problematic consequences gives rise to common interests that prescribe rules and measures to mitigate those consequences, and officials are appointed to enforce them.

Dewey's account of democracy, like Addams's, is an extension of social ethics. However, it also highlights the importance of a public consciousness to be able to understand and respond to harmful "indirect consequences": the results of others' acts. In Dewey's view, modern society has developed and has been structured to obstruct public consciousness. He argues that "the machine age has so enormously expanded, multiplied, intensified and complicated the scope of the indirect consequences, [has] formed such immense and consolidated unions in action on an impersonal rather than a community basis, that the resultant public cannot identify and distinguish itself."[26]

Apathy for public affairs stems from the inability to identify oneself with the issues. The world is an increasingly complex place, which makes it difficult to understand public issues. Mass communication also makes for a distraction from public affairs. The "variety, and cheapness

---

23   Dewey 2016, 153.
24   Dewey 2016, 66.
25   Dewey 2016, 78.
26   Dewey 2016, 157.

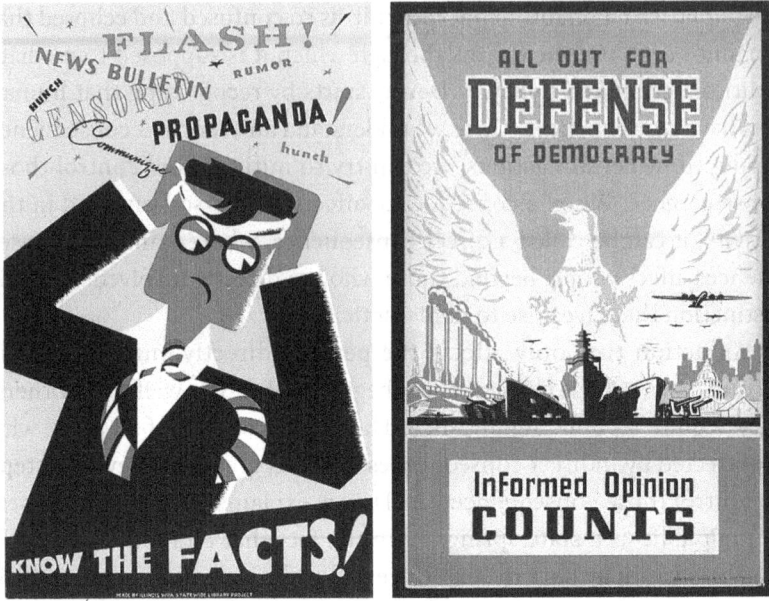

FIGURE 26 · During Dewey and Lippmann's time, it was a common concern that propaganda in the form of news media could undermine democracy and that an informed public was necessary. What is distinctive about the problem we face today with AI?

of amusements represents a powerful diversion from political concern. The members of an inchoate public have too many ways of enjoyment, as well as work, to give much thought to organization into an effective public." He adds, "the movie, radio, cheap reading matter and the motor car with all they stand for have come to stay. That they did not originate in deliberate desire to divert attention from political interests does not lessen their effectiveness."[27]

Dewey was writing in 1927, when mass communication was in its infancy. Today AI enables social media and the spread of information in a way that would have been unimaginable at the time. Yet, it is also clear that social media and the internet have magnified the number of distractions and amusements that might prevent public engagement (Figure 26). Viral videos, addictive social media, and headlines such as "The 25 craziest TikTok challenges so far – and the ordeals they've caused"

---

27   Dewey 2016, 167.

remind us that the challenges to democracy observed in the 1920s have only magnified today.[28]

Public confusion and reliance on expertise stems from science, which gave rise to the industrialized world, being too isolated from our lives. As Dewey explains,

INQUIRER'S TOOLBOX

9. Are there historical precedents or analogies that can inform how I might understand a problem or a potential solution? In what ways is that precedent helpful or relevant? In what ways is it not?

> Science has revolutionized the conditions under which associated life goes on ... But it is not known in the sense that men understand it ... They do not understand how the change has gone on or how it affects their conduct. Not understanding its 'how,' they cannot use and control its manifestations. They undergo the consequences; they are affected by them. They cannot manage them.[29]

Recall the story of Sarah Wysocki and reconsider Figure 17 from Chapter 4. We undergo the consequences of the conclusions of AI, but the average person has little understanding of what produced those consequences and why, thus making it difficult to consciously control or even fully be aware of those consequences.

The isolation of science from our lives is caused by the division between pure and applied science. Science has largely been confined to furthering the interests of "a possessing and acquisitive class." Science is only applied *to* human life, not *in* human life. Dewey explains: "Application in life would signify that science [becomes] the instrumentality of that common understanding and thorough communication which is the precondition of the existence of a genuine and effective public."[30]

In his historical assessment, Dewey sees this separation of pure and applied science as leading to significant social problems,

> The scientific revolution of the seventeenth century was the precursor of the industrial revolution of the eighteenth and nineteenth. In consequence, man has suffered the impact of an

---

28   Cost et al. 2023.
29   Dewey 2016, 189.
30   Dewey 2016, 196–97.

> enormously enlarged control of physical energies without any corresponding ability to control himself and his own affairs. Knowledge divided against itself, a science to whose incompleteness is added an artificial spirit, has played its part in generating enslavement of men, women, and children in factories in which they are animated machines to tend to inanimate machines. It has maintained sordid slums, flurried and discontent careers, grinding poverty and luxurious wealth, brutal exploitation of nature and man in times of peace and high explosives and noxious gases in times of war ... The glorification of 'pure' science under such conditions is a rationalization of escape; it marks a construction of an asylum of refuge, a shirking of responsibility,[31]

This highlights the fact that these problems are not only political, but also moral.

The flaws of democracy result from a failure of social norms to adapt to new conditions. We undergo the consequences of social change but are unable to manage them. What is needed to transform a confused public into what Dewey calls "The Great Community" is education that would enable the public to develop critical methods of judgement. This requires both a freedom to pursue social inquiry and the publicity of its results. Anything that obstructs publicity limits and distorts public thinking.[32] As Dewey explains, until bias, misrepresentation and propaganda, and ignorance are replaced by inquiry and publicity, we have no way of telling how apt for social judgements the populace might be.[33] Education and communication to foster inquiry into social problems so that all might understand the consequences of certain kinds of actions will help create a public consciousness and thus are necessary for a functioning democracy. As Dewey argues, "The clear consciousness of a communal life, in all its implications, constitutes the idea of democracy."[34]

When we compare the accounts of democracy found in Lippmann and Dewey, we see a common recognition of problems. Modern industrial society is distracting and complicated, making it difficult for any

---

31   Dewey 2016, 197.
32   Dewey 2016, 190.
33   Dewey 2016, 226.
34   Dewey 2016, 176.

coherent public to govern itself. Whereas Lippmann believes this is natural, Dewey considers this as signifying major challenges to democratic life. When we think about this debate today and of the potential threat to democracy posed by AI, we also find that some of these challenges are not new. Nevertheless, the debate gives us a framework in which to consider how AI might threaten democracy.

2. AI and the Public Consciousness

The debate between Lippmann and Dewey gives us a sense of the ways we might start to consider the relationship between democracy and AI, and it will help us evaluate whether there could be potential threats to democracy. Does artificial intelligence undermine the public's ability to govern itself? To begin, let's consider the example of AI-powered microtargeting.

2.1 MICROTARGETING

Microtargeting is a form of advertising that utilizes machine learning to allow a political campaign to identify specific sub-groups of voters and to target them with specific kinds of advertising that would resonate with such groups. It uses finely honed messages targeted at narrow categories of voters based on data analysis using individuals' characteristics and consumer and lifestyle habits.[35] It harnesses science for the power of advertising, and it is the latest development of a process that has been decades in the making, in which traditional campaigning has given way to marketing and behavioural science.[36]

Machine learning is simply the newest application of this technology. Using methods like cookie-based tracking, your browsing history can be tracked to target you with specific personalized content.[37] Political campaigns can now easily access information such as names, addresses, gender, and ethnicity through official voter registration lists. They can cross-reference this information with information sold to them by corporations that provide such services to create enhanced voter files,

---

35 Gorton 2016, 62.
36 O'Neil 2016, 187.
37 Bontridder and Poullet, 2021, e32–34.

including up to 900 data points per individual. These may include shopping history, occupation, and education. A model is then built correlating voting behaviour and political views with these variables.[38] By identifying individual voters that they are more likely to convince, a political campaign can match its message to the specific interests and vulnerabilities of the voter.[39]

Cambridge Analytica was a consulting firm that acquired and used personal data from online users to build psychological profiles for political campaigns. They used information from more than 40 million voters, ranking them according to different personality types and developing individual profiles. Each profile evaluated them as potential volunteers, voters, or donors, and reflected their stances on different issues.[40]

Building profiles of individual voters based on their interests can allow a campaign to engage in more targeted advertising. For example, a campaign might attempt to target specific niche audiences such as fathers aged 35 to 44 in Texas who frequent gun-enthusiast sites.[41] In 2016, the Donald Trump campaign targeted audiences of the program *The Walking Dead* with ads about immigration after data analysis revealed that it was more popular with Republicans than Democrats, potentially owing to the show's themes involving a fear of strangers, the prevalence of guns, and rugged individualism.[42]

Microtargeting does not require AI, but applying machine learning to large datasets of voter information does allow a campaign to work on a much larger scale to target more voters more efficiently. Hence there are potential democratic benefits of microtargeting. Microtargeting could increase political participation by allowing politicians to engage voters through more relevant advertisements when more general advertising might not interest them.[43] Exposing such citizens to such political information might increase the chances that they will become interested and vote.[44] It can also allow campaigns to learn more about what is important to voters, particularly subsets of voters that aren't typically heard from.

---

38   Barocas 2012, 32.
39   Borgesius et al. 2018, 82.
40   O'Neil 2016, 191.
41   Borgesius et al. 2018, 83.
42   Saletta, n.d.
43   Barocas 2012, 31.
44   Borgesius et al. 2018, 85.

Microtargeting can also make it cheaper and more efficient to campaign. This could create new opportunities for small and new parties without their having to spend money on expensive TV advertising. Microtargeting can also make voters more knowledgeable about issues that matter to them. Mass communication requires that political parties focus on broad messaging around core issues; however, a targeted approach can allow more political issues to be discussed. Also, as Walter Lippmann pointed out, the sheer volume of information about policies and the available political options can be overwhelming. But microtargeting can expose voters to information most relevant to their voting decisions. Thus, "topics which are only relevant to small audiences may get a market stand in the marketplace of ideas."[45]

But there are potential privacy drawbacks. AI-powered microtargeting requires massive datasets on users to build a model. Some of this data might be sensitive by itself or because it enables the prediction of sensitive preferences of individual voters. Cambridge Analytica acquired personal data from Facebook users under false pretenses. The data of over 87 million users, most of whom had not given Cambridge Analytica permission to access their data, was harvested and used to assist the 2016 campaigns of Donald Trump and Ted Cruz.[46] There is also the possibility that the concern people have for their own privacy will make them provide false information, avoid voter registration, and refrain from political activity.[47] (For further discussion of issues relating to privacy, see Chapter 6.)

Another potential drawback of microtargeting is that it can create an unlevel playing field and potentially exclude certain groups from campaigning. We mentioned above that microtargeting may be beneficially inexpensive. But when it requires a large amount of data and processing power, it can be quite expensive, thus helping to consolidate power for larger, more well-financed parties, while smaller parties will struggle to be competitive.[48] Microtargeting allows campaigns to reach the voters that they are mostly likely to resonate with with surgical precision. However, this also means that parties can ignore those they think unlikely to support them or anyone who doesn't vote. The more a particular group

---

45　Borgesius et al. 2018, 86.
46　Confessore 2018.
47　Barocas 2012, 34.
48　Borgesius et al. 2018, 88.

is considered irrelevant, the more they will be seen as irrelevant, creating a feedback loop. Alternatively, a party could even target their opponent's voters and attempt to suppress their vote.[49]

A significant potential drawback of microtargeting is the effect on public conversation during election campaigns. Microtargeting essentially allows politicians to choose their voters by allowing them to privately target an individual voter with whatever message they think the voter wants to hear. A politician might know that a particular voter dislikes immigrants and so shows this voter personalized ads that suggest the politician wants to curtail immigration. They might then target a different voter who is more open to immigration with ads that suggest they want to help immigrants in the job market.[50]

As Cathy O'Neil explains, neighbours living on the same street might be exposed to very different campaigns: "It will become harder to access the political messages our neighbours are seeing—and as a result, to understand why they believe what they do."[51] This allows political campaigns to manipulate voters. For example, a candidate might be more likely to press divisive wedge issues knowing that most of this campaign material will not be publicly seen. Feeding voters with only certain kinds of information can give them a biased understanding of the major issues in a campaign.[52]

A former member of the Clinton administration warned about the dangers, explaining that "the nightmare scenario is that the databases create puppet masters ... Every voter will get a tailored message."[53] The personalization of campaigning may lead to the "fragmentation of the marketplace of ideas" where voters are increasingly exposed to particular issues they care about but not others.[54] While there is evidence to suggest that people do seek opinions that are different from theirs when it comes to issues that have personal salience, the concern is that they won't hear about other issues. Voters might hear opposing views, but not competing concerns about other social priorities.[55] If people aren't interested in overarching issues because they only focus on issues relevant

---

49 Borgesius et al. 2018, 87.
50 Borgesius et al. 2018, 88.
51 O'Neil 2016, 195.
52 Barocas 2012, 33.
53 Barocas 2012, 31.
54 Borgesius et al. 2018, 89.
55 Barocas 2012, 34.

to them, public debates become less democratic, and the deliberative process is cut short.[56]

Political campaigns can select which voters are most likely to deliver them a victory without having a clear coherent platform. Moreover, to what degree is it possible to claim a clear mandate when victory is achieved based on an incoherent coalition of voters all voting according to personally salient issues? As Cathy O'Neil explains,

> The result of these subterranean campaigns is a dangerous imbalance. The political marketers maintain deep dossiers on us, feed us a trickle of information, and measure how we respond to it. But we're kept in the dark about what our neighbors are fed ... This asymmetry of information prevents the various parties from joining forces—which is precisely the point of a democratic government.[57]

Does AI-powered microtargeting represent a threat to democracy? Microtargeting might encourage people who might otherwise be uninformed and uninterested, and provide the opportunity to participate by providing them with messages that are the most important and most relevant to them. Voters who might have too little time can now more easily access relevant information, offsetting some of the concerns expressed by Lippmann. While they might not enable an omnicompetent citizen, microtargeting might at least make voters slightly more competent, at least on specific issues.

Alternatively, Dewey might take issue with the private way in which microtargeting is delivered and its effect on public conservation. As he argues,

> We have the physical tools of communication as never before. The thoughts and aspirations congruous with them are not communicated, and hence are not common. Without such communication the public will remain shadowy and formless ... Till the Great Society is converted into a Great Community, the Public will remain in eclipse.[58]

---

56   Borgesius et al. 2018, 89.
57   O'Neil 2016, 195.
58   Dewey 2016, 190.

Given that microtargeting has the potential to fragment the public consciousness about overarching issues and that politicians can preach different messages to different voters in a private way, we can see why microtargeting would be a threat to Dewey's conception of democracy.

Alternatively, Lippmann would not likely consider this a threat. The notion of an informed public debate and that the public will act based on that discussion is simply more phantom thinking. As Lippmann argues, the justification for majority rule is in the "sheer necessity" of finding a place in society for force that resides in the weight of numbers.[59] Public opinion is not a "conserving or creating force directing society to clearly conceived ends."[60] Thus, there is no threat from microtargeting in terms of the lack of a public conservation preventing public action.

2.2 BIAS AMPLIFICATION AND POLARIZATION

Historically, there have been relatively few sources for news. Print newspaper, radio, and television were expensive to produce, and this meant that few competitors needed to aim for mass appeal. With the rise of the internet, the number of news organizations has skyrocketed, allowing for news organizations to focus only on specific kinds of issues or to only consider world events from certain perspectives. This makes it much easier to only expose yourself to the kind of news and information that you already want to hear.

Confirmation bias is a tendency to search for, favor, and accept information that supports our existing beliefs, including a tendency to ignore facts that don't fit those beliefs. This can include cherry-picking usable data and stopping an inquiry once we obtain our preferred answer. It can also involve framing questions in specific ways, or describing facts in certain ways to fit a narrative. When you are exposed to news and information online, particularly on social media, algorithms determine what gets shown to people, how likely those things will be shown, and why certain things aren't shown.

Sometimes we might encounter people who seem to have a completely different understanding of the world than we do, as if they aren't getting the same news as yourself. They probably aren't. A Google search,

---

59   Lippmann 1993, 48.
60   Lippmann 1993, 55.

for example, will generate different results by considering information such as your location, age, gender, education, marital status, race, etc.[61] If you get news from social media, your newsfeed will also likely reflect the kind of content that you engage with the most. This is a function of a model that is generating predictions about what kinds of content it thinks you want to see.

The business model of social media companies is to make money by selling information about you to advertisers. This creates an incentive to have you engage with as much content as possible (see Chapter 6). They record every online action to propose content that will optimize the amount of time you spend on the platform.[62] However, your browsing history might still offer too little data for a model to make reliable news recommendations to keep you engaged. To produce reliable recommendations, your data is combined with that of other users that the model believes are similar to you. This is a process called **collaborative filtering**.[63]

When collaborative filtering, a model will compare your interests to the interests of other users. If you share an interest with certain other groups of people and they have an interest in something else, then you might share that interest as well. By comparing similar tastes across different individuals, an algorithm can cluster them into groups and start making new recommendations to individuals who haven't previously expressed interest in a topic. Thus, by looking at your previous search history, by building a user-profile based on your previous browsing habits and personal information, and by using collaborative filtering, an algorithm can generate a highly personalized news feed that provides information it believes you want to see.

Processes like this can create feedback loops. The more you browse certain kinds of content, the more the algorithm recommends more of it to you, thus making you more likely to engage with it. By contrast, content you are less likely to look at becomes less likely to be recommended to you.

This can lead to confirmation bias in multiple ways. First, it means that it will amplify your personal biases since you are only consuming content the algorithm thinks you want to see. Second, it also amplifies

---

61   Shaffer 2019, 35.
62   Bontridder and Poullet 2021, e32–34.
63   Shaffer 2019, 36.

biases of people who are like you since their preferences will influence what is recommended to you. This means, for example, that if a person belongs to a politically extreme group, their personal biases and those of people like them are amplified since collectively their recommendations are likely to filter out more moderate and objective content that might check those biases. They see less content that challenges their thinking, and hence so do you. The biases of both the group and of yourself are reinforced. This can create a filter bubble or echo-chamber-like effect where the only content you engage with is content that supports your worldview.

Data scientist Kris Shaffer explains: "I see things in accordance with my bias, I share a subset of that content that is chosen in accordance with that bias, and that feeds into the biased content the people in my network consume, from which they choose a subset in accordance with their bias to share with me, and so on."[64] Attempting to check this filter bubble by looking at a wider range of news yourself or by befriending a wider group of people on social media will have limited effect because your content will still be recommended via collaborative filtering using other people's habits.

This can not only shield ideas from critical thought, but at the population level it can help contribute to **polarization**. As biases get amplified, a filter bubble will promote uncritical in-group thinking and can lead to vilification of the opposition. Divisive content often performs better, increasing the odds that the algorithm will promote more of the content to keep you engaged.[65] This often magnifies anger towards those who are outside of the group.[66] Over time the public becomes more polarized around what groups think and there is less room for common ground and common understandings of political problems. As Noémi Bontridder and Yves Poullet explain, "As individuals increasingly interact solely with groups of people with their own views, sometimes inflated by social bots, and are manipulated with a biased version of reality, their ability to accept the presence of other cultures and to understand them

---

64   Shaffer 2019, 40.
65   Bontridder and Poullet 2021, e32–35.
66   Shaffer 2019, 41.

is made harder."[67] This is problematic because recent studies have shown that about half of Americans get their news from social media.[68]

Do these processes harm democracy? Walter Lippmann would probably not think so. Even in the early twentieth century, he considered it unreasonable to expect a well-informed public. In a 1919 article in *The Atlantic*, he wrote that no one can even pretend to keep track of all ongoing affairs: "What men who make the study of politics a vocation cannot do, the man who has an hour a day for newspapers and talk cannot possibly hope to do so. He must seize catchwords and headlines or nothing."[69]

Lippmann argued that news was already complicated and confusing, and in drawing a comparison to a trial, he argued that the public makes for a poor jury about what the facts are:

> To this jury any testimony is submitted ... by any anonymous person, with no test of reliability, no test of credibility, and no penalty for perjury ... If I lie to a million readers in a matter involving war and peace, I can lie my head off ... all this for the simple reason that the public, when it is dependent on testimony and protected by no rules of evidence, can only act on the excitement of its pugnacities and its hopes.

If Lippmann is right, then AI wouldn't seem to be able to make things worse.

Reliance on news and reporting from second-hand reports and eye-witness accounts and the selective choices of editors already undermines objective truth in news. As Lippmann explains,

> The news of the day as it reaches the newspaper office is an incredible medley of fact, propaganda, rumor, suspicion, clues, hopes, fears, and the task of selecting and ordering that news is one of the truly sacred and priestly offices in a democracy ... where all news comes at second-hand, where all the testimony is uncertain, men cease to respond to truths, and respond simply to opinions. The environment in which they act is not the

---

67   Bontridder and Poullet 2021, e32–37.
68   Walker and Matsa 2021.
69   Lippmann 1919.

realities themselves, but the pseudo-environment of reports, rumors, and guesses.

Since the question is not what happened, but whether whatever the reporters and editors happened to think, the public is deprived of trustworthy means of knowing what is really going on and people will naturally start to believe whatever news fits their preconceptions. So not only is it not realistic to expect that people will understand affairs that don't concern them, but the roots of the problems of news and public understanding of issues that are relevant to filter bubbles existed long before AI.

Dewey responded that a rapidly industrializing world and the rise of mass communication disintegrated smaller communities without allowing for the rise of a coherent public. As he explains, "There are too many publics, for conjoint actions which have indirect, serious and enduring consequences are multitudinous beyond comparison, and each one of them crosses the others and generates its own group of persons especially affected with little to hold these different publics together into an integrated whole."[70] Thus, for Dewey the most significant democratic reforms will be those that allow such a scattered public to recognize itself and its common problems, and express its interests.[71]

Dewey argues that citizens should have their lives enriched by associating with others. Cutting ourselves off from objective understandings of the world limits our ability to understand it and act. "We" and "our" only exist when the consequences of action are perceived and become an object of desire and effort.[72] Democracy is supposed to empower the individual to form and direct the activities of groups to which they belong to liberate our potential as a member of a group in harmony with common interests. However, "since every individual is a member of many groups, this specification cannot be fulfilled except when different groups interact flexibly and fully in connection with other groups ... democracy is not an alternative to other principles of associated life. It is the idea of community life itself."[73]

---

70   Dewey 2016, 166.
71   Dewey 2016, 174.
72   Dewey 2016, 178.
73   Dewey 2016, 175.

Microtargeting and filter bubbles threaten democracy and represent an ethical problem to the degree they undermine our capacity for sympathetic understanding. If we only focus on messages that resonate with us, if we lack information about issues affecting other people, we cannot identify with their problems and they cannot identify with ours. As Jane Addams explains, "the identification with the common lot which is the essential idea of Democracy becomes the source and expression of social ethics."[74]

3. AI and Democratic Manipulation

Microtargeting is also a form of advertising. Like all advertising, it aims to modify your behaviour. Manipulating you to do something like buying a soft drink with advertising isn't usually considered unethical, but what exactly is manipulation and when does it become ethically problematic? Can AI be used to manipulate people and, if so, does this threaten democracy?

Let's define **manipulation** and identify why it can be morally problematic. Manipulation typically involves an attempt to influence someone else's beliefs or actions by bypassing a person's normal reasoning process and substituting poor reasoning or irrationality. The problem with providing a single definition is that it can be difficult to encapsulate cases that seem to undermine our reasoning in different ways or to determine to what degree the manipulator intends to deceive. But according to philosopher Thomas Christiano, manipulation is primarily an ethical concept that at the very least involves the intentional exploitation of the flawed reasoning of others.[75]

In other words, manipulation involves a person influencing another person to believe or do something by a flawed process of reasoning that the manipulator knows works and which the person being manipulated is unaware of and unable to rectify. This could include using false information or emotional manipulation such that the manipulator knows the emotion will interfere with reasoning. Taking advantage of one's rational capacities means that if that person were to approach the issue with

---

74 Addams 1902, 11.
75 Christiano 2021, 3.

more time and without flaws in reasoning, they would likely arrive at a different conclusion or action than the manipulator intended.

Christiano argues that from a democratic standpoint, there are three problems with manipulation. First, it deprives a person of their ability to think on their own and reach a conclusion that is justified. Second, it substitutes one person's aims and beliefs with another's. Lastly, he argues (as we have already seen) that manipulation can undermine the democratic aim of getting people to appreciate other people's interests. As he explains, manipulation is a threat to political equality: "[It] arises here when there is a systematic sorting of persons into groups of persons engaged in manipulation and those effectively subjected to it. Such a systematic sorting implies the setback of the interests of members of the subject group in favor of the interests of the perpetrators."[76] With a firmer understanding of manipulation and some of the ethical harms associated with it, let's consider how AI can be used to manipulate others.

3.1 DISINFORMATION

AI can threaten democracy by spreading **disinformation**. Whereas misinformation is incorrect or misleading information, disinformation aims to deceive and confuse others about the facts. We've already discussed how collaborative filtering and algorithms can target you with content that an advertiser believes is more likely to keep you engaged. This business incentive means that an algorithm can spread disinformation to millions of people with great efficiency. It also creates opportunities for users themselves to spread it to others.[77]

AI can spread human-generated disinformation by targeting those who are most vulnerable to such messaging, but it can also generate its own disinformation. Malicious stakeholders can create social bots to disseminate information. For example, Russia is a well-known sponsor of such bots (machines posing as people online) with which it attempts to manipulate social media for its own purposes and to interfere with elections.[78] At one point there were over 23 million of these bots on Twitter alone.[79] They can post context-relevant content to blend in before

---

76   Christiano 2021, 4.
77   Shaffer 2019, 40.
78   Menn 2023.
79   Hajli et al. 2022, 1238.

posting disinformation designed to cause political strife, skew online discourse, and manipulate markets.[80]

AI can generate disinformation using **deepfakes**. Deepfakes rely on generative adversarial networks, where one network generates an image and tries to fool another network into thinking the image is real. Using hundreds or thousands of photos of someone or bits of speech, it can generate a convincing new visual image of that person or fake speech.[81] Political campaigns have already begun to use deepfakes. An Indian politician named Manoj Tiwari used deepfakes in a campaign where the original English version of his speech was translated and read by an actor in a dialect of Hindi with Tiwari's face overlaid on the actor's face.[82] Voters who watched those videos might have voted differently had they realized the screen image wasn't really Tiwari.

Deepfakes can spread disinformation in many ways. For example, they might feature depictions of real people saying or doing things they would never say or do in order to discredit them or to incite unrest. Deepfakes also present an opportunity to intimidate, blackmail, and sabotage political candidates, or give them the opportunity to plausibly deny any negative media attention they have had.[83] For example, an Indian politician named Palanivel Thiagarajan claimed that a politically embarrassing recording leaked to the press was the product of AI.[84] Lies about a candidate during an election can be spread, discrediting that candidate's policies and making them less viable. Further still, a campaign might spread disinformation about voting and voter registration to suppress voter turnout. The spread of disinformation can also be a double-edged sword in that not only can disinformation mislead but it can undermine the credibility of legitimate information by causing doubts about its veracity and shattering trust in institutions.[85] (Further discussion of deepfakes can be found in Chapter 6.)

---

80 Bontridder and Poullet 2021, e32–35.
81 Bontridder and Poullet 2021, e32–33.
82 Lyons 2020.
83 Kertysova 2018, 67.
84 Nilesh 2023.
85 Bontridder and Poullet 2021, e32–34.

3.2 HYPERNUDGING

Nudging involves instances where a choice is presented or structured using positive reinforcement and indirect suggestion to alter a person's behaviour in a predictable way without restricting options or significantly changing incentives.[86] If a choice is given to people to be an organ donor where the default option is changed from not being an organ donor to being one, with the possibility of opting out, more people will likely be organ donors.[87] The person is nudged into making a certain decision because it is the default option. The nudge won't affect everyone, especially those who are consciously opposed to an organ donor plan, but it does make a difference to people who might not use a rational process to decide. This creates an opening to manipulate the choice to ensure that people will select a certain outcome by taking advantage of that irrational process.

Starting in the 1990s, some airports put a picture of a fly in men's urinals to encourage men to "hit the target." The result was a reduction in spillage by 80%.[88] Stores in London discovered that rioters would be less likely to attack their store if they painted their outside murals with the faces of babies. A speed bump might count as a form of nudging to encourage drivers to slow down. Some districts have taken the concept further by putting optical illusions on the road that from a distance resemble speed bumps.

Nudging isn't inherently wrong, but it becomes manipulative when the aims of the person being nudged and the person doing the nudging are at cross-purposes and the nudge takes advantage of flaws in rational capabilities to achieve an end the person being judged would not endorse if they were more rational.[89] It isn't wrong to create a system that encourages you to do something you would have done anyway, or something that you would have done if you had enough time to think and reflect. But if the nudge works on the assumption that the person being nudged won't have access to certain information, or relies on an assumed confirmation bias, or that people won't have time to make well-reasoned choices, then the nudge becomes manipulative.

---

86   Thaler and Sunstein 2008, 6.
87   Christiano 2021, 4.
88   Hooker 2017.
89   Christiano 2021, 5.

Hypernudging uses AI to continuously reconfigure the nudge based on data that an algorithm is getting from you and others.[90] When you shop online, the site might have an algorithm that generates additional recommendations for purchases according to what you or people like you tend to buy. The system can continuously adjust itself with every new purchase. In politics a group might take advantage of hypernudging to encourage certain groups of people to vote. For example, in the 2010 and 2012 US elections, Facebook conducted experiments by adding "I voted" updates to people's newsfeeds to create greater peer pressure to vote.[91]

Nudging won't work on those who aren't particularly interested in what they are being nudged into doing. The purpose of nudging is to slightly increase the probability that I will do something. But when hypernudging can take advantage of AI, it can operate using massive scales of data and reach massive numbers of people. A 0.001% increase in the probability that someone will do something may not sound like much, but applied to millions of people it can start to make a huge difference in behaviour.[92] Potentially, it could even change an election result.

3.3 STATE ACTOR MANIPULATION

Governments also have the potential to use AI to manipulate and undermine democratic processes. Certain governments have been known to engage in operations to attempt to influence public opinion by spreading propaganda or disinformation on social media. Governments can now use AI to accomplish this on a larger scale.[93] These efforts might include a government attempting to manipulate public opinion in its own country or in another nation.

Governments can use AI to undermine democracy by engaging in surveillance and utilizing facial recognition. Such technology has the capacity to violate privacy by identifying people who may wish to remain anonymous, making it easier, for example, to identify protestors and activists who criticize the government. We've discussed facial recognition bias, but governments could also attempt to use facial recognition to surveil and control their populations. In America, for example, police

---

90   Yeung 2017, 119.
91   O'Neil 2016, 180.
92   Christiano 2021, 7.
93   Silverberg 2023.

have used facial recognition to identify protestors.[94] In the UK, the police fine people for covering their faces from facial recognition cameras.[95] Amazon announced that they would temporarily halt development of facial recognition technology for government use following complaints from the American Civil Liberties Union and numerous other civil rights organizations.[96]

In addition to surveillance, government actors can use AI to spread disinformation and undermine democratic rule. In Myanmar, Facebook has been widely used and was often confused with the internet itself. Almost half of the country's population use Facebook and most get their news from that site.[97] Taking advantage of all this, the Myanmar military used their operatives to create false accounts to spread incendiary messages targeting the country's Muslim population with disinformation. Facebook's own investigation later revealed "clear and deliberate attempts to covertly spread propaganda that were directly linked to the Myanmar military."[98]

In 2016, a genocide began against the Rohingya Muslims in Myanmar including the murder and rape of tens of thousands of people and the burning of villages. The result was a refugee crisis where nearly 690,000 people were forced to leave their homes.[99] Part of the military's success in spreading propaganda was owing to Facebook and its failure to police its own policies. As Facebook whistleblower Frances Haugen revealed, about 87% of Facebook's budget for classifying misinformation was spent on the United States, leaving only 13% for the rest of the world, even though the US comprises only 10% of users. In her view this meant that Facebook was "literally fanning ethnic violence."[100]

3.4 INFORMATIONAL POWER

According to Christiano, a key component of democracy is the ability to participate in democratic deliberation, something he calls "informational power." It has two dimensions: the first involves the ability

---

94   Vincent 2020.
95   Dearden 2019.
96   Solon 2020.
97   Akinwotu 2021.
98   Mozur 2018.
99   Al Jazeera 2018.
100 Popli 2021.

to understand information and seek out information about how to advance one's interests along with the common good, while the second is the ability to disseminate information.[101] Without the ability to understand what is in your own good, voting power is of little use. But this already suggests potential for manipulation, as AI offers an extraordinary amount of informational power in that an algorithm can determine what you see and how likely you are to see it.

Using algorithms in this way requires funding and since funding is likely to come from the wealthy, using algorithms for the purpose of disseminating information will be carried out with the interest of the wealthy in mind, thus suggesting a possible conflict of interest between sources of wealth that disseminate information and those that receive it.[102] Like Lippmann, Christiano recognizes that people have a limited budget of time and energy to absorb political information. He argues that this is what makes them susceptible to manipulation through microtargeting and hypernudging.[103]

We depend on what Christiano calls a "cognitive division of labor," whereby much of the knowledge and information we rely on to make decisions for ourselves is held by others. For example, we rely on qualified experts like doctors or mechanics to make decisions based on the information they give us, even if we don't have all the information ourselves. This creates a cognitive dependence on others, Christiano explains, "because our cognitive abilities are not very developed to take up the slack when those who we might depend on are no longer performing their roles as they ought. And cognitive abilities are quite weak when the social environment on which we depend is not, epistemically speaking, reliable."[104]

The difference in epistemic conditions means that some groups will be better placed to understand and advance their interests within a political system than others. One of the consequences of groups having less access to information and a lesser ability to discriminate between the quality of different pieces of information will be greater vulnerability to messages that are attention grabbing but have little epistemic merit.[105]

---

101 Christiano 2021, 9.
102 Christiano 2021, 9.
103 Christiano 2021, 8.
104 Christiano 2021, 10.
105 Christiano 2021, 12.

This makes such people highly susceptible to manipulation using algorithms. This can include the use of emotionally charged messages or the use of misleading information. Manipulation can take place if someone takes advantage of the fact that others suffer from epistemic weaknesses and vulnerabilities caused by the social environment.

Using algorithmic communication, groups with more money can manipulate the information provided to everyone else, take advantage of those who are more cognitively dependent, and exploit their cognitive weaknesses. Does the cognitive dependence of certain groups on information provided to them by algorithms open them to manipulation that is ethically not justifiable from a democratic standpoint?

Some of the scholars discussed clearly have concerns about AI manipulating democracy given their conception of what makes for a strong democracy. Christiano believes that democracy requires that people share equally in the effective political power over the society they live in: "To the extent that the epistemic weaknesses generated by a particular environment contribute to many people being deprived of important tools for thinking about their interests or aims, and vulnerable to having their rational abilities subverted by others with different purposes and interests, we have a serious problem for democracy."[106] Describing the threat of disinformation, Noémi Bontridder and Yves Poullet claim that, "when individuals' capacity to participate fully in public debate is impaired, including through the effective manipulation of their opinion and voting decisions, democracy is seriously jeopardized."[107]

But why worry about people being deprived of information that they might not be interested in and wouldn't understand anyway? What does it even mean to participate *fully* in public debate? Lippmann might think these concerns are premature. The public can never be expected to be reasonably informed about anything; most people don't participate much in public affairs either, so is there really a novel threat from AI? Recall that Lippmann did not believe a single "public" even exists. The daily business of running the world and dealing with substantive and technical issues falls on those directly involved with those issues. As he notes, "When problems arise, the ideal is a settlement by the particular interests involved. They alone know what the trouble really is."[108]

---

106 Christiano 2021, 12.
107 Bontridder and Poullet, 2021, e32–36.
108 Lippmann 1993, 63.

The general public is mostly unaware of these issues and if they are aware of them, their understanding will always be limited because they are outsiders to the issue. The role of public opinion "is to align men during the crisis of a problem in such a way as to favor the action of those individuals who may be able to compose the crisis."[109] Public opinion, as Lippmann calls it, is merely "a reserve of force" brought in during a crisis to defend those who would find workable social rules and to prevent the use of arbitrary, brute, and unaccountable power; "by cancelling lawless power [public opinion] may establish the condition under which law can be made."

Although the public will not be truly able to understand the issues they must grapple with, Lippmann argues that during a full free debate, advocates on various sides might expose one another's real motives: "Open debate may lead to no conclusion or throw no light whatever on the problem or its answer, but it will tend to betray the partisan and the advocate."[110] Thus, by exposing those who only seek to act in their own self-interest in arbitrary and unaccountable ways, the public might align with a side that seeks a workable solution. However, Lippmann never presumes that public debate will reveal the truth.

From Lippmann's perspective, it doesn't seem like AI manipulation necessarily generates a distinctive threat to democracy. Algorithms can generate and spread propaganda, but propaganda was widespread in Lippmann's day as well. The notion that the public could even be well-informed or that every person should have a chance to "fully participate" in democratic life are artifacts of romanticized notions of democracy. If AI were to cause us to misunderstand the world or make people less engaged with all the diversity of issues that typically fall under the category of "public affairs," it would not be a new development for democracy.

Is there any threat that AI might pose to democracy according to Lippmann's account? Perhaps collaborative filtering has the potential to inhibit the function of public debate, namely, to identify and expose partisans. In so doing, collaborative filtering is more likely to exclude from your attention debates you aren't interested in, rather than exclude both sides of debates that you are interested in. Perhaps Lippmann would

---

109   Lippmann 1993, 58.
110   Lippmann 1993, 104.

consider the opaque nature of AI to be the biggest threat to the function of democracy. Recall that, "It is the function of public opinion to check the use of power in a crisis, so that men, driven to make terms, may live and let live."[111] But when AI is opaque, it creates the potential for the use of arbitrary and unaccountable forms of power that the public will be unable to check.

Would Dewey consider the potential of AI to manipulate public opinion a threat to democracy? Mass communication now, just as in the 1920s, tends towards sensationalism and manipulation. As Dewey notes, publicity, as it existed then, largely meant, "advertising, propaganda, invasion of private life, the 'featuring' of passing incidents in a way which violates all the moving logic of continuity, and which leaves us with those isolated intrusions and shocks which are the essence of 'sensations.'"[112] Today, media and news feeds encourage people to consume as much online content as possible. Often issues of public importance come to our attention but are soon forgotten. Social media such as TikTok, YouTube shorts, and Instagram reels are powered by algorithms that incentivize content to be short and limited in scope and depth. This enables public manipulation as well.

As Dewey explains, "Emotional habituations and intellectual habitudes on the part of the mass of men create the conditions of which exploiters of sentiment and opinion only take advantage. [Humans] have got used to an experimental method in physical and technical matters. They are still afraid of it in human concerns."[113] Once again the threat that emerges is related to a split between science and society. It is only when the potential consequences of science are understood in terms of how they relate to our lives that we can try to better manage the consequences of scientific progress. A society can only become a Great Community when "the ever-expanding and intricately ramifying consequences of associated activities shall be known in the full sense of that work, so that an organized, articulate Public comes into being."[114]

Part of this process will involve instilling what Dewey calls a "scientific attitude" into our daily lives. According to Dewey, the consequences of scientific progress are determining the relations in which human

---

111   Lippmann 1993, 64.
112   Dewey 2016, 192.
113   Dewey 2016, 192.
114   Dewey 2016, 204.

beings sustain each other. But "if it is incapable of developing moral techniques which will also determine these relations, the split in modern culture goes so deep that not only democracy but all civilized values are doomed. A culture which permits science to destroy traditional values but which distrusts its power to create new ones is a culture which is destroying itself."[115]

Adopting a scientific attitude entails a disposition towards inquiry rather than acting according to routine, prejudice, unexamined tradition, or self-interest.[116] As Dewey explains, "While it would be absurd to believe it desirable or possible for every one to become a scientist when science is defined from the side of subject matter, the future of democracy is allied with the spread of the scientific attitude. It is the sole guarantee against wholesale misleading by propaganda."[117] In so far as AI has the potential to undermine social inquiry, it has the potential to undermine the public.

While Dewey and Lippmann may have disagreed about whether democracies were capable of aspiring to something better, they understood that mass communication and new technology could manipulate and disorient the public. AI isn't necessary for targeted advertising or to create and spread propaganda, and in that regard the potential threats to democracy are not new. Does AI pose any distinctive threats to democracy? Cathy O'Neil notes that it is scale that turns problematic models into major problems.[118] When considering the threats to democracy from AI, computer scientist Ognjen Arandjelovic explains, "we are not dealing with any sui generis aspects of AI but rather with a change in scale; quantity rather than quality."[119] Christiano also confirms that "the distinctive character of algorithmic communication is the sheer scale of the data on which the algorithms can operate."[120] The threats to democracy posed by AI are not novel in kind, but AI has the potential to exponentially make the long-standing tension between science and democracy far worse.

---

115 Dewey 1989, 118.
116 Dea and Silk 2019, 350.
117 Dewey 1989, 114.
118 O'Neil 2016, 30.
119 Arandjelovic 2021, 8.
120 Christiano 2021, 6.

4. Suggested Solutions?

If microtargeting, collaborative filtering, disinformation, hypernudging, and cognitive dependence are problems for democracy, what can be done? China introduced regulations to make sure that algorithms are not designed to optimize the time we spend engaging online, but instead are optimized for our well-being.[121] These include provisions that ensure that people can turn off recommendations which can prevent the creation of algorithms designed to addict. They prohibit algorithms that create false news or information. They are designed so that people can be aware of the mechanisms behind an algorithm and to promote an understanding of the basis of an algorithm's recommendation.

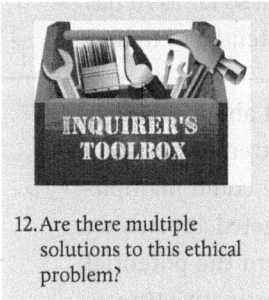

INQUIRER'S TOOLBOX

12. Are there multiple solutions to this ethical problem?

But we must also accept the fact that it may not be easy to regulate or prevent the use of methods such as political microtargeting or hypernudging. These, like any other form of political advertising, could be considered political speech protected under the rights to freedom of speech. Nevertheless, it is possible that some aspects of microtargeting could be regulated, for example, by forcing political campaigns to disclose how much they spend on such practices or by creating a public registry where any political advertising must be stored for future reference.[122]

Christiano argues that a first step towards fixing issues relating to collaborative filtering and cognitive dependence is to introduce regulations requiring that platforms use algorithms that ensure there is a diversity of opinion in your feeds, such as the Fairness Doctrine in the United States from the 1950s until the 1980s.[123] He also advocates for more education for those who are worse off epistemically.

Perhaps the answer is simply more democracy in the form of industrialized democracy. This means a corporation is accountable to its employees, who are given a vote in terms of how the company is run. Many people who have worked for tech companies, for example, have come forward to blow the whistle on what their company was doing because they were aware of the addictive and destructive tendencies

---

121 Romero 2022.
122 Borgesius et al. 2018, 93.
123 Christiano 2021, 13.

their algorithm could produce. Is it possible that if Facebook were democratically accountable to its employees, the company wouldn't have allowed the kind of democratic manipulation that took place in Myanmar? Or, would those employees still be so self-interested that even a democratically run Facebook would tolerate such things?

What if the public itself controlled some of these algorithms? In the twentieth century, mass media and advertising led many to worry about the state of democracy and this gave rise to public broadcasters such as the British Broadcasting Corporation, the Canadian Broadcasting Corporation, and the Public Broadcasting Service. Is there a similar twenty-first-century equivalent? The creation of publicly owned social media, for example, might negate the economic incentive to create addictive algorithms. The public would also have greater control over how the algorithms that run the service operate, increasing the amount of public accountability for the kind of news and information that one is exposed to. On the other hand, such services could themselves be sources of manipulation by state actors in the form of censorship, the invasion of privacy, and propaganda.

5. Conclusion

The use of AI to generate and spread misinformation or to target you and others to manipulate behaviour on a large scale makes the prospect of coherent democratic governance more difficult. Additionally, we are living in an increasingly technical world where we need to rely on experts to govern our own affairs. Is AI too complicated for the public to understand? To what degree is it reasonable to expect that the public will understand AI and its ethical consequences?

Is democracy even up to the challenge? Some have argued that in the face of public ignorance about most issues, increasingly polarized tribalistic politics, and an increasingly complex world, democracy should be replaced by some form of epistocracy, or expert rule.[124] Philosopher Jason Brennan has argued that given the high-stakes nature of many public decisions, citizens have a right to competent administration. But given that the public is largely incompetent at make ruling decisions,

---

124 Runciman 2018.

given their ignorance and irrationality, some form of expert rule should replace equal voting power.[125] Ognjen Arandjelovic has advocated that we should use AI to aid in the transition of governments towards a mixture of democratic and non-democratic elements.[126]

As we've discussed in this chapter, the prospect of transferring the responsibility for decision making from the public to experts is problematic. As Lippmann notes, experts are themselves largely outsiders to problems and are not always fully competent or impartial. Dewey also makes the point that expertise is not divorced from values and ethics. We might hope that experts will rule in the best interests of everyone, but translating expertise into practical action requires political judgement, which is not the same thing as expertise.

In this chapter we have considered the potential threats that artificial intelligence poses to democracy from the perspective of a more idealistic and perhaps cynical account of democracy. Even if you find yourself ultimately disagreeing with both perspectives, these contrasting views give us a sense of the possible ethical concerns involved if a democracy deteriorates in the face of technological change. Both views might find the lack of transparency and accountability a threat to democratic order. Both perspectives recognize that the invasion of privacy and the manipulation of the public at large with mass communication are not new phenomena but have been part of an ongoing process of societal adaptation to scientific and technological discoveries. Even though the rate and scale of change of these discoveries has only magnified, the central ethical question of the relationship between science and society remains largely unsolved. What is new is the scale at which AI can operate and the lack of transparency inherent in most algorithms that make these problems worse.

---

125 Brennan 2011, 708.
126 Arandjelovic 2021, 8.

ADDITIONAL MATERIAL

Democracy • a political system that holds that the authority and legitimacy to act depends on the consent of the governed.

Sympathetic understanding • learning different perspectives and experiences to develop empathy and form a cooperative response to a moral problem.

Microtargeting • a form of advertising that uses big data and machine learning to identify groups that are susceptible to certain messaging and attempts to alter their behaviour.

Confirmation bias • a tendency to search for, favour, and accept information that confirms or supports one's existing beliefs.

Collaborative filtering • a process that feeds users what they will like, based on the reactions of similar users.

Polarization • the practice of dividing people along ideological, political, or other lines, thus reducing the space for compromise, empathy and cooperation.

Manipulation • deliberately influencing, altering, or distorting someone's understanding or perspective on an issue by bypassing or taking advantage of weaknesses in someone's rational processes.

Disinformation • false information that is designed to mislead people.

Deepfake • a video of someone where the face is digitally altered so that they appear to be someone else using AI.

Hypernudging • deliberately influencing, altering, or distorting someone's understanding or perspective on an issue.

---

1. What is the relationship between Addams's concept of sympathetic understanding and the process of democracy as a social ethics?

2. What can democracies do to protect themselves from the biggest threats from AI?

3. What is confirmation bias? What are some examples of it that you see in your own life or in your beliefs? How might we overcome or limit such a bias?

4. What is the meaning of disinformation, and how does it differ from misinformation?

5. Many people worry about deepfakes and about not being able to identify cleverly made ones. What are some of the central issues here? Offer examples.

6. Are you more inclined to agree with Dewey's account of democracy and its ills, or with Lippmann's? Is there a middle view? How do different understandings of democracy inform our understanding of the impact of AI on democracy?

6

AI and Humanity

So far, we've mostly discussed ethical issues involving AI from the perspective of how the models AI produces are created, the biases they can produce, the consequences of their error, and their lack of transparency. However, we haven't considered in depth how AI can affect humans themselves. We've inquired into what impact AI can have on democracy, but that inquiry only begins to suggest the range of possible ethical issues involving how AI can affect our behaviour for better or for worse and considering the consequences of artificial intelligence for human beings and human life.

How can an algorithm affect your behaviour? Is it okay if it does and, if so, should algorithms be designed to make us better or less selfish? Can an algorithm addict you or undermine your sense of self? Can it manipulate your feelings? Can it undermine your health? If AI can generate false information, can it also undermine our capacity to learn and understand our world? If someone creates a deepfake of your face or a model that recreates your voice, is this an invasion of your privacy? This chapter will examine the ethical issues involved when humanity must adapt to developments in artificial intelligence. It will consider these central questions:

1. How did the concept of privacy develop as an ethical concept?

2. What are the different meanings of privacy? Which meanings matter the most?

3. How does the attention economy give rise to surveillance capitalism?

4. What is addiction? Can an algorithm addict someone?

5. What are the ethical problems associated with using artificial intelligence to create art?

6. Can deepfakes and other AI-generated content undermine our ability to know the world?

7. Should algorithms be designed to change our behaviour? What ethical concerns are involved when people attempt to manipulate an algorithm for their own ends?

8. How does the interaction of AI and humans give rise to potential ethical concerns?

Using what we've learned in previous chapters, we will apply our inquirer's toolbox and examine various cases in terms of existing ends-means relationships and the types of ethically salient consequences produced, as well as consider the ethically salient assumptions involved.

1. Privacy

When considering privacy issues related to AI, there are two groups of morally salient consequences to think about. The first are privacy concerns stemming from the use of algorithms. For example, an algorithm might be able to predict very personal information about you, or someone might create a deepfake of your face, or a government might use facial recognition to track your every move. The second group of consequences are those that follow from the growing incentive of companies to collect as much information about people and their habits as possible for the purpose of training data (Figure 27). Not only are there more groups out there trying to collect information about you, but this information can be hacked and potentially reveal your personal information. To investigate the ethics involved, let's examine what privacy is.

The ethics of privacy might seem straightforward. People have a right to privacy in the sense that people have a right to private lives free from public scrutiny. We might even be tempted to claim that it is ethically

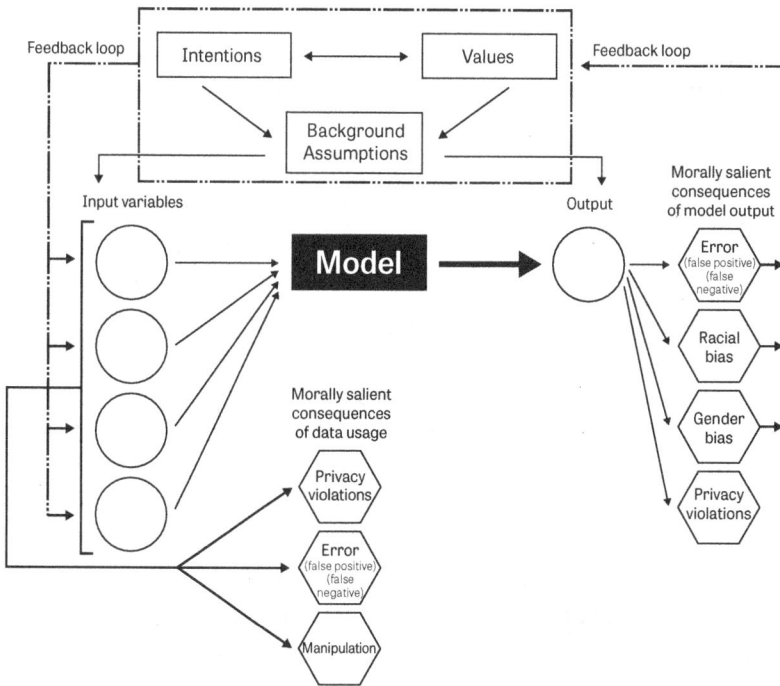

FIGURE 27 · Some privacy violations follow from the use of an algorithm. A model might reveal personal information or capture your face in a deepfake. Other privacy violations follow from the collection and storage of massive amounts of personal information necessary for training those algorithms. Nefarious agents can always hack databases and correlate information between different datasets to learn private information about you.

wrong for someone to invade someone else's privacy. But when we consider some practical cases, we find that the issue of privacy is very complicated. Consider the following situation. You need to open your bag for airport security: is this an invasion of your privacy? You might say yes, but given that this is an everyday occurrence that most people would accept, let's assume it isn't. Why not?

Given that an airport has a vested public interest in the safety of air travel, perhaps there is no reasonable expectation of privacy in those circumstances. But what if airport security requires you to provide the password for your phone or laptop: is that an invasion of privacy? Now you don't just have to open your bag, you must provide confidential information. You might think that this is a step too far and you shouldn't have to provide that information. But let's compare these two cases. In both instances, the authorities are asking similar questions and for similar

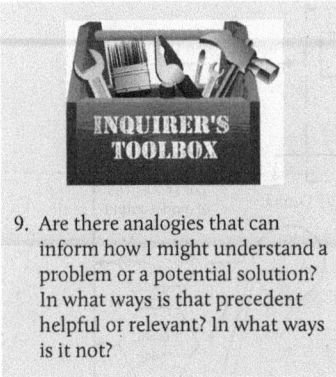

reasons; so what is the moral difference between being forced to open your bag and being forced to provide a password, particularly if we continue to hold the assumption that an airport is inherently not a private place? Are there different levels of privacy and does it matter where we are? If so, why?

What if your roommate were to enter your room, open your personal journal, and read it: would that be a violation of your privacy? We might think that this is an obvious yes because it's a personal journal and it wasn't left out in the open to be read. What if someone goes through your garbage, which is sitting in a public alley, and happens to discover that you have a medical condition? We might be inclined to think that this too is an invasion of privacy. But what if you leave your personal journal in the kitchen and your roommate inadvertently reads some of it? We might be inclined to be more forgiving in that case since the journal was left out in the open and the roommate didn't intend to learn the information. But if that's the case, what's the moral difference between that and the person who goes through your trash? The information was in a public place and the persons involved weren't necessarily looking for your personal information, yet we might have two different intuitions about these cases.

Imagine Sherlock Holmes is able to look at you and infer everything about you. He observes your clothing and deduces what your occupation is. He observes your body language and can tell you have a mental health condition. He looks at what you're carrying and infers that you're having family problems. We might think that that kind of information is private, yet Holmes can infer such things using information that is public and out in the open. Is this an invasion of privacy? Of course, Holmes doesn't actually *know* these private things about you; he infers them as the most probable explanation based on what he has observed of you, which is public and in the open. Nevertheless, we might still feel that that is privileged information even if it is just an educated guess. But is it ethically wrong to make inferences like that if the information is publicly available?

If we say that Sherlock Holmes did nothing wrong by making such inferences, what about an algorithm that can read your tweets and infer your political, social, and ethical beliefs from them, even if you haven't posted them? Would that be ethically problematic? What is the moral difference between Holmes and the algorithm? If it is problematic, where do we draw the line when we say it is an invasion of privacy? Also, do we bear any responsibility for the information we put into the public that might allow others to learn private things about us? If it isn't ethically wrong, then why would it be wrong for someone to go through your garbage? What is the morally relevant difference?

As we can see by considering some practical cases, privacy is not as straightforward as we might think. We might be inclined to expect privacy in some cases but not others and for different reasons, so we must be cautious when making sweeping ethical claims about privacy. It's easy to say that someone shouldn't invade someone else's privacy, but where do we draw the line when it comes to information you put out in the world for others to discover? What if someone could learn something very private about you based on seemingly unrelated information that you put in public? Let's consider the historical origins of the concept of privacy and what ethical functions we think privacy should perform in our lives so that we can consider the issue in relation to AI.

1.1 THE ORIGINS AND FUNCTION OF PRIVACY

The distinction between a public and private sphere of action dates to the classical age, when philosophers like Aristotle drew a political distinction between the *poli* (public) sphere and the *oikos* (private) sphere.[1] Modern conceptions of privacy, however, were largely informed by developments in the nineteenth and twentieth centuries.

In the nineteenth century, J.S. Mill articulated the harm principle: "The only purpose for which power can be rightfully exercised over any member of a civilized community, against his will, is to prevent harm to others."[2] This principle establishes a means of protecting private actions and beliefs from public regulation provided they don't harm others. Mill's principle is also a defence of free speech from what he called the "tyranny of the majority," establishing the principle that private people should have a right to their own private thoughts and beliefs. It was also

---

1   Aristotle 2013, 247.
2   Mill 1978, 9.

around this time that the concept of property rights expanded to protect not only physical property but also intellectual property in the form of copyright and patents. Not only was land no longer considered part of a shared commons, but ideas were protected from the public domain.

In 1890, Samuel D. Warren and Louis Brandeis wrote an article for the *Harvard Law Review* titled "The Right to Privacy" which was seminal in articulating contemporary conceptions of privacy. They believed that over hundreds of years of English and American common law, the principle of a right to privacy had developed that wasn't covered by other legal concepts. They argued that "Political, social, and economic changes entail the recognition of new rights, and the common law in its eternal youth, grows to meet the demands of society."[3] The concept of privacy was a living and evolving concept that changed over time in response to new social conditions. They argued that changing technological developments had required a new understanding of privacy: the right to be left alone expressed through the ability to prevent the publication of private materials. As they explained, "Instantaneous photographs and newspaper enterprise have invaded the sacred precincts of private and domestic life; and numerous mechanical devices threaten to make good the prediction that 'what is whispered in the closet shall be proclaimed from the house-tops.'"[4]

In other words, the concept of privacy largely developed with technological and scientific progress. In the twentieth century, it developed to prevent public interference over private affairs. In the United States, the right to access contraceptives and abortion were justified by the right to privacy. In the decision of *Griswold v. Connecticut* in 1965, Supreme Court Justice William O. Douglas asked, "Would we allow the police to search the sacred precincts of marital bedrooms for telltale signs of contraceptives? The very idea is repulsive to the notions of privacy surrounding the marriage relationship."[5] This precedent was key in later decisions regarding abortion rights and same-sex marriage.

Historically, two broad concepts of privacy have emerged. The first can be expressed as a right to freedom from government interference in one's private life and private affairs. People have a right to make personal decisions about themselves and their families. The other conception of

---

3   Warren and Brandeis 1890, 193.
4   Warren and Brandeis 1890, 195.
5   *Griswold v. Connecticut*, 381 U.S. 479 (1965).

privacy is that one has a right to be left alone, in the sense that people have a private life that is free from public attention and public scrutiny. People have a right to control access to their own personal information and to prevent public disclosure of private facts. Nevertheless, this doesn't really tell us what privacy is or what its limits are. It's worth noting that the definition of the concept will follow when we already have a sense of what is private and what isn't and what will count as an invasion of privacy. Thus, we must take care not to attempt to define the concept merely based on our own preconceived notions to justify what we already think should count as private.

INQUIRER'S TOOLBOX

9. Are there historical precedents that can inform how I might understand a problem or a potential solution? In what ways is that precedent helpful or relevant? In what ways is it not?
11. Are there biases or limitations on my perspective which might require insight from others? Am I cherry-picking information, cases, principles, theories, solutions?

1.2 MEANINGS OF PRIVACY?

Privacy might be understood in terms of information control or, in other words, the ability to control when, how, and to what extent information about us is communicated with others. For example, philosopher William Parent notes that defining privacy is difficult owing to the "inconsistencies, ambiguities, and paradoxes" in our language, but nevertheless defines privacy as "the condition of not having undocumented personal knowledge about one possessed by others."[6] However, one problem with this definition is that Parent believes that if that information is already a matter of public record, then you have no right to privacy. So, for example, if your employer were to look you up on social media and see your public posts and keep tabs on your social media, this would be entirely justified.

This could be problematic in terms of AI. In 2012, a man walked into a Target angry that the store was sending his daughter coupons for pregnancy-related items, even though she was still in high school. It turns out, however, that Target was using a model that connected customers' names, demographic information, and purchase history to make

---

6   Parent 1983, 269.

recommendations.[7] It suggested that there were strong correlations between certain purchases (such as unscented lotions) and being pregnant. The man later apologized when he found out that his daughter was indeed pregnant. In this case a company was able to infer private information based on publicly available information to learn something that even the family wasn't aware of.

AI permits additional similar kinds of applications where a model can take publicly available facts in the world and infer information that people might wish to be kept private. For example, an algorithm developed for Twitter by a company called Hunch was able to look at who followed you on Twitter and who you followed, and use that information to predict personal information about you. The algorithm could predict with about 85% accuracy your positions on same-sex marriage, abortion, what kind of media you watch, your spending habits, and more, even if this information was not public. We might say that an algorithm that is able to predict personal information like this is an invasion of privacy, but Parent's definition doesn't capture this sentiment.

Let's consider a different definition. According to Stanley Benn, "A general principle of privacy might be grounded on the more general principle of respect for persons ... To conceive someone as a person is to see him as actually or potentially a chooser, as one attempting to steer his own course through the world."[8] Whereas Parent might argue that once information becomes public record, there is no reasonable expectation of privacy, Benn might argue that observing us when we don't wish to be, or when we are unaware of it, changes and undermines our understanding of our own life experience, and of our self-respect. Anything that undermines us in this way is wrong unless we consent to being observed or recorded, regardless of whether the information is out in the open or not. These are only two definitions of privacy, but already we can tell that different definitions will yield different answers, so perhaps it is worth considering how we might want to define privacy with a mind to the practical cases that artificial intelligence will present.

---

7   Hill 2012.
8   Benn 2010, 228–29.

1.3 SCOPE OF PRIVACY?

Even once we have a working definition of privacy, we must examine its scope. Consider the case of employment "wellness" programs in which employers require that employees submit medical and health-related information to them to qualify for employee benefits, otherwise they will face a penalty. Some corporations require that employees reach a certain number of daily points by providing more information, or that employees maintain certain health metrics, otherwise they have to pay a fee. This information can be sold to third-party companies who can use that data to build models to help market products.[9] Not only does this mean that your employer can monitor your activities in your private life, but also that this information can be sold for profit. Is this a violation of your privacy? What if you consent? Can one really consent if you are only doing so under duress, and otherwise resent the intrusion?

What about if someone creates a deepfake of a celebrity using the thousands of photos of that person widely available online? Is it a violation of privacy to recreate someone's face without permission, even if it relies entirely on widely and publicly available photos? What about, as we've discussed in Chapter 5, if governments use cameras combined with facial recognition to monitor public places? Is there no such thing as privacy in a public place?

Much of what makes questions like these tricky is trying to determine the limits of the public and private spheres. There are occasions where the public might have a right to regulate certain private behaviours if they are in the "public interest." For example, is getting vaccinated a personal private health choice or does the potential risk to public health suggest that it isn't merely private? Questions like this suggest that there is no clear context-independent line when it comes to privacy, in terms of what takes place in the home compared to outside of it. A religious person who believes that a nation was founded to be Christian society, for example, might believe in privacy, but more naturally think that issues like contraception or reproduction inherently fall within the public sphere.

Philosopher Judith Jarvis Thomson argues that such cases reveal how difficult it is to argue them as a matter of privacy. The concept isn't clear enough to allow us to ethically distinguish between cases where we think

---

9   Hancock 2015.

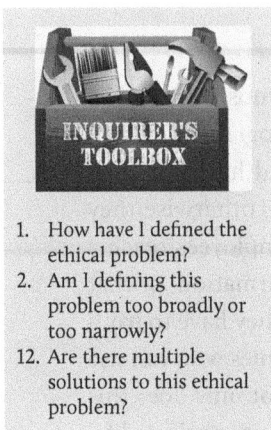

INQUIRER'S TOOLBOX

1. How have I defined the ethical problem?
2. Am I defining this problem too broadly or too narrowly?
12. Are there multiple solutions to this ethical problem?

a violation of privacy has taken place and where it hasn't.[10] She argues, for example, that it is wrong to defend a right to abortion as a matter of privacy. Abortion, like many other issues that are commonly associated with privacy, are better justified as a matter of property rights than privacy rights, such that we understand our own bodies as being property which we have a right to control.[11] In other words, privacy and a right to privacy are ambiguous concepts, and it isn't always clear what kinds of ethical issues fall within the scope of the concept of privacy. Alternatively, ethical disputes about privacy might not come down to the question of whether a certain area should be considered private at all. There may be agreement that while certain kinds of things are generally private, there might be an overwhelming public interest in the case.

1.4 DOES THE CONCEPT OF PRIVACY REQUIRE REFINEMENT?

Given the historical origins and the complex cases we've discussed, some are skeptical about whether there is a right to privacy and even if such a right can be coherently defined. Nevertheless, the concept has developed according to changing social circumstances and social problems. If privacy is an ethical concept used to help understand ethical problems, what is its value? Warren and Brandeis defended the value of privacy based on an "inviolate personality."[12] This means that the essence of a human being, including our individual dignity, integrity, and personal autonomy and independence, is respected. Privacy can also be valuable in that it is necessary for the formation of intimate relationships of love, trust, and respect.

Even if we can't easily define privacy and its parameters, it is a socially and ethically important concept. Yet, as we've alluded to, novel social conditions create new problems that require critical thinking about the concept and what purpose we want it to have in moral discourse.

---

10   Thomson 1975, 296.
11   Thomson 1975, 303.
12   Warren and Brandeis 1890, 205.

AI requires massive data to produce reliable algorithms. Thus, there is a strong incentive to collect personal data. In turn, algorithms can be very effective at predicting potentially very private information about you which can then be used to market things to you. This has created financial incentives to learn as much information about the habits of individuals as possible and AI allows those who seek to profit from this information to operate on a large scale and in new ways. How should we understand the concept of privacy in the age of AI?

Datasets present privacy concerns because the information could be hacked or stolen. To prevent this, many organizations anonymize their datasets. Nevertheless, even if the data is anonymous, it is relatively easy to cross-reference the information with other information known about a person to infer who they are. For example, to help create a better recommendation algorithm, Netflix released datasets containing movie ratings by their customers. The data contained a unique but meaningless ID number for each customer and data about their movie ratings. Even though no other information was released, if someone was aware of the approximate dates when a person had reviewed a few movies, this information could be cross-referenced with other databases online to uniquely identify people 99% of the time. Even revealing simple movie ratings can be an invasion of privacy in that it can reveal things like political leanings and sexual orientation.[13]

To prevent anonymous records being cross-referenced with other information to identify unique personal records, one solution is to introduce **k-anonymity**. The idea is to redact information from the dataset so that no single data record is unique. Imagine, for example, that you were looking at anonymous hospital records and were trying to identify patients. There are no names, but they do include information such as age, postal codes, and gender. If we know that Sally was a 35-year-old female who lived in a certain region, and there is only one anonymous record of a 35-year-old female who lives in that region in the dataset, we can infer that the record belongs to Sally. The purpose of k-anonymity is to withhold enough information so that there are no unique records that can be easily correlated with other information to infer the identity. A dataset can be said to be k-anonymous if any

---

13   Kearns and Roth 2020, 25.

combination of attributes in the database matches at least k number of individuals.[14]

K-anonymity might sound promising to preserve privacy given that no single person can be identified while preserving the utility of the data. However, as Kearns and Roth argue, k-anonymity is problematic. While it prevents us from identifying the record of any one person, it allows us to identify clusters of records that still fit with what we are looking for, narrowing down the options. If, for example, you were trying to identify a specific person and three of the records fit what information you did know about that person, you have a one-in-three chance of identifying the record. If this can be cross-referenced to information from other databases, and even if they too are k-anonymous, it can help identify personal records.

Kearns and Roth argue that **differential privacy** is a better alternative. Differential privacy holds that someone shouldn't be more likely to learn anything about you that could affect you, whether your data is included in the dataset compared to whether it is not. In other words, no single record of a given individual should affect the probability of the outcome very much, so in effect there is nothing that someone can learn about any given individual in the dataset.[15] Differential privacy works by adding statistical random noise to the data to minimize the importance of any one record. When looking at aggregate results, because we know how the noise was introduced, we can later compensate and remove the noise but maintain privacy.

This works because the noise added to individual records means they can't be considered reliable, yet the aggregate results are. The noise could be added by the data scientists themselves in what is called central differential privacy, or it can be added at the local level by the individual subjects though local differential privacy. In the local case, the process would work using randomized responses. Imagine that before answering a question, the data subject essentially flips a coin. If it is heads, they answer honestly, and if it is tails, they provide a random answer, again using a coin flip. This introduces a degree of **plausible deniability** in that any one record isn't reliable, yet the aggregate results are.

---

14   Kearns and Roth 2020, 28.
15   Kearns and Roth 2020, 37.

Differential privacy can be dialed higher or lower by increasing or decreasing the amount of noise depending on the privacy concerns. However, greater privacy protection in the form of greater noise means that more data is required. Error introduced by randomness will shrink to zero the more data is included. In other words, in differential privacy there is a trade-off between privacy and accuracy.[16] Obtaining greater data can often be more expensive and can include privacy concerns in the collection, so it may not always be ethically feasible to maximize privacy. As we alluded to earlier, privacy sometimes is only one value among several that relate to an issue.

K-anonymity and differential privacy are technical definitions of the concept that data scientists and statisticians can use to translate the concept more easily into practical action in terms of datasets. But are these useful developments of the concept? Is privacy in the form of plausible deniability good enough? It can protect the security of a dataset, but it won't protect you from algorithms predicting personal information about you from correlations of known data. There are also potential risks in that with centralized differential privacy, the actual records still exist, and we have enough trust in the data scientists to properly obscure the data. It also won't obscure every record the way that local differential privacy will.

We might question whether differential privacy really is privacy at all. Even in a decentralized process where every record has a built-in level of plausible deniability, we can still make reasonable assumptions about the likelihood that those records are accurate. And, even if they aren't accurate, there is still a problem. Assume someone spreads personal lies about you. People don't actually know something truly personal about you, but they believe that they do. Even though their beliefs are false, this still seems like an invasion of privacy. Thus, even if a record has plausible deniability, someone might still think that they have learned personal information about you which can cause just as many ethical harms as if the information were true, even if you do deny it.

---

16   Kearns and Roth 2020, 44.

2. Social Media Addiction

In the previous section we discussed the incentives of corporations and AI developers to learn as much about you as possible. Algorithms that recommend content and advertising will only be effective if there are large amounts of personal data that track what you do online. In turn, creating a model from your behaviour can allow a corporation to advertise something to you in just the right way to get you to make a purchase, and the more time you spend online, the more you can be advertised to. We do not pay for most social media services online. Social media sites like Facebook, Instagram, TikTok, and X (formerly Twitter) primarily make money by selling advertisements and using their algorithms to learn about you and market products most likely to appeal to you. In such business models, you are not the customer: the advertiser is. You (or rather your attention) are the product. This is what is called the **attention economy**.

The fact that your attention is the primary commodity means that social media companies are incentivized to get your attention, even to manipulate you to get you to give up as much of your time as possible. Before we consider how it does this, it's worth taking the time for some personal reflection. How many hours of social media do you consume in a day? Do you ever find that you consume more than you intend? Does social media ever cause you to neglect your responsibilities? Do you ever try to cut back and find you can't? If you answered yes to any of these questions, you may be addicted to social media. But how is social media designed to be addictive?

2.1 ADDICTIVE DESIGN

A significant influence on the design of addictive algorithms comes from the Stanford Behavior Design Lab. It was founded in the 1990s by psychologist B.J. Fogg to study **persuasive technology**, or the study of how computers can change the way that humans think and act. Fogg has been called the "millionaire maker" for the influence he had on his students. By 2007 he was running courses at Stanford on app design where key figures who would later work for Google, Facebook, and Twitter learned to design applications that could pull in millions of users.[17] Fogg's model

---

17   Basen 2018.

of behaviour understands behaviour in terms of our motivations, our ability, and certain triggers.

The use of intermittent variable rewards is a key feature of making a design addictive. Instead of providing consistent predictable feedback or content, it varies the feedback. We experience hikes in dopamine when we anticipate something, and this can make us want to keep doing it. For example, an app might cut off the bottom half of the bottom row of images on a screen. These images offer a glimpse of what is to come if the user keeps scrolling. As the user pulls the screen, more images load (including new partial images at the bottom). This pull-to-refresh design creates a slot machine effect where the user can continue to play and get fresh rewards.[18] Similarly, a social media app will store notifications and will then provide them to you in a timed way to keep you engaged, or it might take a few seconds to load content to keep you in anticipation.

In addition to intermittent rewards, the pull-to-refresh function is designed to erode natural stopping cues. In sites like TikTok, YouTube Shorts, or Instagram Reels users must continually scroll without end. Without a stopping cue, such as the end of a page, the user never naturally comes to a decision to continue or not. The less opportunity to reflect on your action, the more likely you are to continue scrolling. Another key design feature of addictive design is the use of social validation and reciprocity. The use of features such as "like" buttons incentivizes giving and receiving social validation from others, and social media is designed in such a way as to exploit your desire for social validation by making you seek likes or comments. The use of photo tagging is another way that an app can get you to give it attention by seeking social validation.[19]

AI can refine these tactics in a way that is tailor-made for you. These models can be tweaked experimentally to figure out how to elicit certain responses from you. Several studies have suggested that significant portions of the population use social media almost constantly and have difficulties giving it up. As early as 2011, a study revealed that 59% of Americans reported feeling addicted to social media.[20] A 2022 Pew Research Poll found that 35% of American teenagers are on social media almost constantly, and 54% reported that it would be hard for them to

---

18   Wu 2016, 187.
19   Bhargava and Velasquez 2020, 7.
20   Cabral 2011, 9.

give up social media.[21] A study done the same year found that a third of social media usage was caused by self-control issues.[22] But can you really become addicted to social media?

2.2 WHAT IS ADDICTION?

For most of its existence, addiction has been understood as a substance issue. The idea that behaviours can be addictive has been more controversial, with the American Psychiatric Association not officially recognizing gambling addiction until 2013. Can someone be addicted to shopping? What about video games or sexual intercourse? Can you become addicted to social media just as you can to drugs or gambling? Would it be morally wrong to create such a product? Let's consider what an addiction is and why it can be morally problematic. There is no single definition of addiction. Some theories of **addiction** hold that addiction is a choice, others that addiction is a kind of disease, others that addiction is a learned behaviour, and still others that addiction is a product of neurobiology.

Within medicine itself there are different diagnostic standards. The *Diagnostic and Statistical Manual of Mental Disorders*, 5th edition (DSM-5), defines addiction in terms of eleven diagnostic criteria. Some criteria include whether you use longer than intended, if you've had trouble cutting back, if you experience intense cravings, if you experience withdrawal from lack of use, and so on.[23] Meanwhile, the International *Statistical Classification of Diseases and Related Health Problems*, 11th revision (ICD-11), uses only seven criteria. Outside of medicine, different definitions will hold that an addiction involves a compulsion, while others will hold that addiction doesn't involve compulsion.

There isn't one clear, definitive sign by which one can say that someone has an addiction. What metrics or markers should you use before you claim that someone is addicted? And how severe should those metrics be? Recall from Chapter 3 that in our discussion of clinical depression, Kristen Intemann pointed out that defining depression involves value judgements about the kinds of things we think are vital to a good human life. So too with addiction, where a

---

21   Vogels, Gelles-Watnick, and Massarat 2022.
22   Allcott, Gentzkow, and Song 2022.
23   American Psychiatric Association 2013.

coherent judgement will involve an appraisal that the behaviour similarly threatens what we think is vital to a good life (such as healthy social relationships). We must also be careful, however, not to cherry-pick a definition or only those criteria that happen to fit a predetermined conclusion about what we happen to think is addictive. For example, a corporation may wish to deny that their AI product is addictive by adopting a definition of addiction that precludes behavioural addiction.

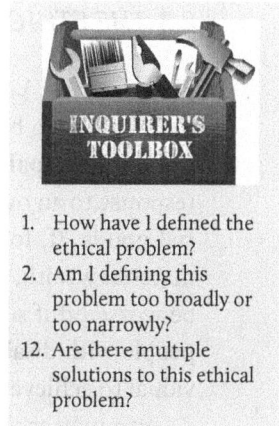

INQUIRER'S TOOLBOX

1. How have I defined the ethical problem?
2. Am I defining this problem too broadly or too narrowly?
12. Are there multiple solutions to this ethical problem?

Depending on the definition, it might not make sense to say that people can be addicted to behaviours. If you adopt the pathology theory of addiction, then does it make sense to say that someone's addiction to the internet demonstrates a pathology about them? Thus, to some extent, to conclude that someone can be addicted to social media is partially a moral judgement.

Vikram R. Bhargava and Manuel Velasquez argue that you can become addicted to social media and that it is morally wrong to create such addictive services. They argue for a conception of addiction that includes six components: it is salient (in other words it occupies most of your thoughts), it modifies your mood by producing a buzz or high, you can build up a tolerance to it though prolonged use (requiring more to produce the same effect), you experience withdrawal if you cannot access the activity, the addiction produces conflicts within the person or those around them, and you are unable able to control yourself and relapse if you attempt to stop.[24] These largely overlap with diagnostic criteria for common substance addiction, and there is evidence that excessive internet use exhibits similar symptoms to other addictions. Functional neuroimaging demonstrates that brains that heavily use social media exhibit similar patterns to brains with substance addiction, and the same molecular pathways that operate in substance addiction can be found in cases of internet addiction.[25]

---

24   Bhargava and Velasquez 2020, 4.
25   Bhargava and Velasquez 2020, 5.

2.3 THE ETHICS OF ADDICTIVE DESIGN

Bhargava and Velasquez argue that making social media addictive is morally wrong because it is harmful and demeaning to its users. They utilize the **capabilities approach** pioneered by Martha Nussbaum in response to an overemphasis in ethics on economic resources and material goods. If, for example, you compared two agents with the same material wealth, one of whom was disabled, the disabled person would be worse off if society affords limited opportunities and capabilities to people with disabilities. What is paramount is the freedom of the individual to achieve.

As a matter of justice, we must consider not only resources but the opportunities that people have to fulfill their capabilities. Nussbaum argues that these capabilities are necessary for "the dignity of the human being, and of a life that is worthy of that dignity."[26] She articulates ten capabilities that are necessary for such a life, of which Bhargava and Velasquez rely on these seven:

1. Life: the ability to live a human life of normal length.
2. Bodily health: the ability to live in good health and proper shelter.
3. Senses, imagination, and thought: the ability to use our senses, to imagine, and to learn, think, and express oneself.
4. Emotions: the ability to have attachment to things and other people.
5. Practical reason: the ability to articulate a conception of the good and critically reflect about one's life.
6. Affiliation: the ability to live with and for other people and to engage in meaningful social interaction.
7. Play: the ability to laugh, play, and enjoy recreational activities.[27]

Their argument is that making social media addictive is ethically wrong because it limits our capabilities. Studies suggest that the more time adolescents spend on social media, the more likely they are to feel unhappy and depressed.[28] If we compare the rates of teen depression and anxiety over time, we find a sharp uptick, particularly among girls after

---

26   Nussbaum 2003, 40.
27   Nussbaum 2003, 41–42.
28   Shakya and Chistakis 2017; Raudsepp and Kais 2019; Kross et al. 2013.

2010. Around the same time there was a similar increase in suicide rates and nonfatal self-inflicted injuries among adolescents in the US.[29] From 2010 to 2014, the rates of hospital administration for self-harm did not increase for women in their twenties, but it doubled for girls aged 10 to 14.[30] The generation who entered middle school when social media came to prominence are less likely to start dating and far more likely to report being lonely at school.[31]

Human beings are adapted to seek social approval from our peers, but we are not adapted to seek it and the dopamine hit it provides from hundreds or thousands of people on a daily basis.[32] People tend to have a worse body image the longer they use social media.[33] As former Facebook executive Chamath Palihapitiya explains,

> We curate our lives around this perceived sense of perfection because we get rewarded in these short-term signals—Hearts and Likes and Thumbs-Up. We conflate that with value and we conflate that with truth, and instead what it really is is fake brittle popularity that's short-term, and leaves you even more vacant and empty.[34]

Former Pinterest President Tim Kendall puts the matter more bluntly: "It's plain as day to me. These services are killing people. And causing people to kill themselves."[35]

People with social media addiction also have significantly less time for other aspects of their lives.[36] The addiction can negatively affect your capacity to engage in play and recreational activities with others. It can also limit your ability to affiliate with others: people with addiction tend to engage in fewer social activities and spend less time with family and friends.[37] The negative mental health effects also impact our emotional states, as those with an addiction are more likely to feel depressed, have

---

29   Miron et al. 2019; Twenge, Martin, and Spitzberg 2019.
30   Mercado et al. 2017.
31   Haidt and Twenge 2021.
32   Reynolds 2018.
33   Haidt 2021.
34   Sulleyman 2017.
35   Allen 2020.
36   Andreassen 2015.
37   Bhargava and Velasquez 2020, 11.

See a former Google engineer discuss the effects of social media on public consciousness.

low self-esteem, feel more social anxiety, feel alienated from family and peers, and are less likely to have interpersonal relationships.[38] Social media addiction can also impact your ability to reason accurately and concentrate and to direct and control your own affairs. So, it negatively affects both your capacity for sensation, imagination, and thought and your capacity for practical reason.

Thus, Bhargava and Velasquez argue that addictive social media is ethically wrong because it harms fundamental capabilities that are necessary for a good life. But why is social media driven by AI morally worse, or even morally different, from the economic forces that finance media like television and radio? The answer is that the algorithm learns and adapts based on your individual behaviour. This fact adds insult to injury because not only does social media addiction harm you, but unlike television, radio, or even actual drugs, the algorithm continually refines the platform itself in order to make it more addictive for each user and to make them engage longer.[39] The more you engage with social media, the more the algorithm learns how to keep you engaged, feeding you exactly what will keep you on longer and creating a feedback loop.

This is ethically problematic because it involves the user themself in the process that makes them more addicted to the platform. As Bhargava and Velasquez argue, this demeans and disrespects the user and conveys the idea that companies do not care if they use the person to harm themself. Further, they argue that this is an ethically wrong form of exploitation.

Social media companies exploit vulnerabilities in the form of the cravings that arise from the addiction and from the pervasiveness of the internet itself in order to advance their own ends.[40] In other words, not only does the addiction create a craving to use that a company can exploit, but because the internet is difficult to avoid in contemporary life, and because we often need it for work or social interaction, we are placed in situations where we are more tempted to use social media, even if we wanted to avoid becoming addicted. Given that social media companies demean and disrespect their users by co-opting them into

---

38   Twenge, Martin, and Campbell 2018.
39   Bhargava and Velasquez 2020, 14.
40   Bhargava and Velasquez 2020, 15.

helping to make the platform more addictive, this constitutes a morally objectionable form of exploitation.

Do we find Bhargava and Velasquez's argument convincing? In terms of ethical inquiries into this matter, any conclusions we draw will depend on the terms of the discussion. As noted, there is room for disagreement about the nature of addiction, what constitutes symptoms of social media addiction, and how severe they must be to constitute addiction. We might object that most research discussed involves only correlations and does not indicate a causal relationship. It is possible, for example, that depressed people are more likely to become addicted to social media. However, longitudinal and experimental studies have also supported the idea that social media is addictive. One study found that the more addicted people were to Facebook, the worse their physical and mental health became over a three-year period.[41] An experimental study found that users who stopped using social media showed a reduction in depression levels while the control group who kept using did not.[42]

In August 2018, a group of sixty psychologists published an open letter to the American Psychological Association calling attention "to the unethical practice of psychologists using hidden manipulation techniques to hook children on social media and video games."[43] Should psychologists be involved in helping to design potentially addictive technology? Tim Kendall believes the product he helped create kills people. Chamath Palihapitiya has expressed "tremendous guilt" over his part in helping to create a service that he believes is ripping society apart.[44] One does not have to build a nuclear bomb to become a destroyer of worlds, and many who have helped create these services have later expressed regret, just as Robert Oppenheimer did. Once again, we must ask, "How ethically responsible are researchers for designing things that harm society on large scale?"

2.4 SUGGESTED SOLUTIONS?

One potential way to improve the situation is to make changes to these services either by getting the service to voluntarily act or through

41   Shakya and Christakis 2017.
42   Hunt et al. 2018.
43   Screen Time Action Network 2018.
44   Vincent 2017.

regulation. For example, social media services could keep track of your usage statistics and warn users about the amount of time they are using. The service could also be more transparent by warning users in advance that the algorithm is designed with your time and attention in mind and that the service is potentially addictive.[45] Should social media come with warning labels just as cigarette packages often do? The services could also be redesigned to make it easier to quit. For example, if one wishes to quit Facebook, the user must go through a convoluted series of menus, they must confirm several times that they wish to close their account, and they are bombarded by messages and photos of friends showing everything the user will miss out on if they quit. Redesigning those features to make it easier to quit might at least help with the problem.

More direct solutions would attempt to disrupt the market dynamics of the attention economy. Social media services would be less inclined to create addictive services if their users were the customers. Services that require paid subscriptions for their revenue, for example, do not need to try to maximize screen time in the same way that a service dependent on ad revenue does. In other words, a subscription-based business model does not require that the platform be addictive as an essential feature of the service.[46] Would it be better if you had to pay to use services like Facebook and Twitter? While it might make the service less addictive, it would also introduce barriers for lower-income individuals to be part of social media. In Chapter 5 we considered the idea of creating publicly owned social media services. This carries certain challenges as well, such as obtaining community agreement on standards of content. Another solution might be to simply ban social media altogether until the ethical issues can be sorted out. This might seem extreme or to place a limit on free speech, but given that we could also think of social media as a giant open-ended human experiment, perhaps there is an argument that social media in this form is more socially harmful than helpful.

---

45   Bhargava and Velasquez 2020, 23.
46   Bhargava and Velasquez 2020, 20.

3. Generative AI

We've discussed the relationship between AI and modelling. We've discussed how models can be used for evaluative purposes such as reviewing job applicants or evaluating medical scans for disease. However, models can also be used for generation. For example, if you sample various speech recordings, you could break those recordings down into various units of sound such as phones (which include any distinctive sound or gesture), phonemes (units of phones that can distinguish one word from another), and diphones (the transitions between two phones) and use that to train a model that will generate new audio speech from the recordings using text prompts. Although generative AI can be used for producing images, sounds, text, and video, the development and use of such AI offers distinctive ethical challenges.

3.1 DEEPFAKES

In Chapter 5, we discussed the use of deepfakes in a democratic context. A deepfake involves the use of AI to manipulate a piece of media to replace one person's likeness with that of another. Just as we can create a model that will produce an audio voice as an output, we can create a model of someone's face using photographs as training data and that model will take other people's faces as an input and will produce the face it was trained on as an output. The technology appeared on the internet in 2016 and has been used on many celebrities and politicians. For example, in 2021 a very realistic deepfake of Tom Cruise was released on TikTok, which went viral and picked up millions of views.[47]

This technology has the potential for many positive applications in art and media. For example, some queer and transgender people have been able to enjoy AI-augmented versions of themselves using AI avatar art.[48] However, it also presents significant ethical concerns. To begin with, there is the issue of consent. If you are a celebrity, for example, there are potentially thousands of photos of you on the internet. It is extremely easy to find publicly accessible images that can be used as training data. Anyone could potentially create a deepfake model of their

---

47 Vincent 2021.
48 Roberts 2022.

face without their consent, as in the case of Tom Cruise. A public figure might have their image used in ways they didn't intend, such as advertising or marketing, or their face could even be used for pornography.[49] As a public figure, you might consent to images of yourself being in public, but where is the ethical line drawn that determines how much of that content you should ultimately control? Does the ethical importance of consent increase if the deepfake is being passed off as authentic or is it equally important regardless?

What if you are not a public figure? While more training data will produce better results, it is possible to create deepfakes with relatively few photos. There are reports of deepfake pornography being generated from ordinary people on the internet by scraping images from social media.[50] If we are tempted to forgive non-pornographic celebrity deepfakes but not deepfakes of private people, is it because the latter haven't put themselves into the public eye in the same way? If so, wouldn't that also imply that if we post publicly available photos on social media, we too shouldn't expect to have full control over how those images are used? Where do we draw the line when it comes to non-public figures?

The idea that public figures must grant consent for every conceivable use of their image is a tall order that may be practically unenforceable. We wouldn't normally think that collecting images of a celebrity and putting them into a photo collage would be unethical unless we had the celebrity's consent. What is the moral difference between that and a deepfake? The difference is what the deepfake allows me to: mislead. If, with the intent to mislead, I use those photos to pass myself off as the celebrity or try to imply that a celebrity is endorsing something they haven't, I have done something wrong. If I use those photos to put celebrities into compromising and embarrassing scenarios, I am demeaning and disrespecting them. Deepfakes are complicated ethically speaking because we could understand the problem in different ways. Consent might be the paramount issue in this case, but it might require resolving some of the issues we

INQUIRER'S TOOLBOX

1. How have I defined the ethical problem?
2. Am I defining this problem too broadly or too narrowly?
12. Are there multiple solutions to this ethical problem?

---

49   White 2022.
50   Hao 2021.

discussed relating to privacy and to the boundary between what is public and what is private. On the other hand, we can also choose to focus on the harms relating to demeaning and deceiving people.[51]

Deepfakes can also spread misinformation. In 2018, Jordan Peele worked with BuzzFeed to create a deepfake of Barack Obama to bring public attention to the potential of deepfakes to spread misinformation.[52] One could produce fake videos of political opponents saying or doing embarrassing things, or videos aimed at confusing voters about where a candidate stands on an issue.

Sometimes mere confusion is all that is necessary to help someone advance their interests. Getting someone to believe something can be difficult, but creating doubt through deception can be easy. Another means of creating confusion is using deepfakes as an opportunity to create plausible deniability about what you might have said and done, blaming it on a deepfake.[53] In Chapter 5, we discussed cases of politicians taking advantage of this, and we will consider further issues surrounding deepfakes in the following sections. However, all of this suggests the risk that public trust in almost any media could eventually disappear.

3.2 ART AND MEDIA

In 2022, video game designer James M. Allen won a digital arts competition at the Colorado State Fair for his piece "Théâtre d'Opéra Spatial." It was controversial because Allen produced the image with AI. The reaction to his win was harsh, with some claiming that it was unfair to use AI in the contest. Some claimed that Allen had won simply by "pressing a few buttons," while others claimed that "if creative jobs aren't safe from machines, then even high-skilled jobs are in danger of becoming obsolete."[54] Was submitting AI-generated art to the contest unfair, as many claimed, and does generative AI represent a threat to the future of art and media?

Allen made his entry using an AI called Midjourney. Using text prompts to describe what the user wants, Midjourney generated images.

51   De Ruiter 2021, 1319.
52   Vincent 2018.
53   Zhang et al. 2020.
54   Gault 2022.

Allen generated over 900 images before choosing a short list of three.[55] He made additional adjustments using Photoshop and boosted the resolution with a tool called Gigapixel.[56] The Colorado State Fair defines their digital arts category as "artistic practice that uses digital technology as part of the creative or presentation process." The work was submitted using the name "Jason M. Allen via Midjourney." Despite Allen's attempt to make it clear that he had used AI, the judges later claimed that they were not aware the piece had been generated by AI, but would have still awarded him first place based on the piece itself.

Was it unfair of Allen to submit his piece and win? The rules did not preclude the submission of AI-generated work, Allen was transparent that the piece was AI-generated, and the judges, in hindsight, still believe it should have won. Some might still argue that it was wrong for Allen to enter the contest because he was not the artist: the artificial intelligence created the work. However, we must be careful how we try to define the nature of this problem. Was the AI the artist in this case or was the AI an artistic tool for Allen, the true creator? While Allen spent his time differently than other artists, it doesn't mean that generating the image took no skill or didn't require the aesthetic taste necessary to choose one out of hundreds of images and still win. The judges claimed that they chose the piece for the impact it had on them, and the themes of the image were a deliberate choice by Allen. The AI itself, on the other hand, had no sense of the impact of its work; it didn't think any differently about the art in the first image than it did in the 900th image.

Generating art using AI requires data to train the models. Thus, when AI produces visual media, it uses images it was trained on, sometimes even replicating the signatures of artists whose work was used as training data (Figure 28). This raises issues of plagiarism. An AI piece is not a direct copy, but it does take "inspiration" from the art that it was trained on. However, art is often about meshing previous works and techniques to create something new, so it is difficult to see this as a mere copy of other artist's work.[57] If we are inclined to think that AI-generated art is not just a copy, and if the AI is not the artist, then it would follow that Allen is the artist. If Allen is not the artist, then the

---

55   Metz 2022.
56   Kuta 2022.
57   Das, Chakraborty, and Arunsaravanakumar 2022.

FIGURE 28 · John Dewey having a conversation with a robot. Produced using SDXL 1.0.
SDXL 1.0 is released under the CreativeML OpenRAIL++-M License. Details on this license can be found at generative-models/model_licenses/LICENSE-SDXL1.0 at main · Stability-AI/generative-models · GitHub.

piece of art would seem to have no artist, and perhaps it shouldn't be considered a piece of art at all.

Given Allen's process and intentions, it would be difficult to argue that this is not art simply because he didn't directly produce the image himself for the same reason that it would be difficult to argue that photographers are not artists because they don't make the image directly themselves. Nevertheless, when cameras were first invented, people believed that this would spell the end of painters.[58] However, AI cannot replace the esthetic choices and judgements of humans because it is we who must decide if anything produced by AI has merit. The role and the tools of the artist might be changing, but their place in whatever system that produces artistic works remains paramount.

Of course, this does not mean that anything generated by AI counts as art. One can, with very little thought or effort, type in a prompt and generate an aesthetically impressive image. However, such images will lack thought about what the image is supposed to express or communicate artistically. Allen did consider these things, and this is what distinguishes Allen's use of the AI from the way anyone else might use it.

Still, we might still consider generative AI to be ethically problematic because of the economic threat to professional artists. Regular

58 Lee 2022.

paying work in the arts can be hard to come by and producing art can be time-consuming and expensive. AI will be able to generate content far cheaper, faster, and more efficiently than human artists. As one artist put it, "This thing wants our jobs, it's actively anti-artist."[59] Not only does AI have the potential to destabilize the art industry by taking jobs, but it adds insult to injury by using artists' works as training data.

Similar problems arise when we think about generative AI and mass media such as movies, television, or music. In addition to being able to generate a static image, AI can be used to make music and, as we've discussed, deepfakes. In a short time, the world has moved from early deepfakes appearing on the internet to the deployment of deepfakes in film and television. Digitally de-aging actors with computer-generated imagery has been around for a few years, but films like *Indiana Jones and the Dial of Destiny* have been able to digitally de-age their stars using deepfake technology.[60] Similarly, deepfakes can be used to bring characters played by deceased actors back to the screen. AI can also be used in the writing process to produce scripts.

In music, AI can be used to generate models of a musician's voice, allowing one to produce audio tracks where an artist's voice can be recreated as if they were really singing. This has the potential to make recording a cheaper and more efficient process for the same reason that Auto-Tune—a program that corrects the pitch of a singer's voice—is heavily favoured by the industry. In 2023, a group was able to create an AI "reunion" of the band Oasis, styled as "Alsis," using AI to add Liam Gallagher's voice to tracks.[61] On YouTube, one can find countless AI-generated vocal tracks that make it seem as if one singer is singing a song they never recorded. Ever wondered what Donald Trump and Joe Biden would sound like recording the 1978 song "Sultans of Swing"? What about Johnny Cash singing "Barbie Girl"? One could use AI to create just about anything.

There is a potentially lucrative marketplace for AI-generated music, but the use of AI models that replicate an artist's voice and appearance presents ethical and legal concerns. Actors and musicians will be under pressure to sign over legal rights to AI replicas of their voices and faces, so that studios can continue to create products without the artist's

---

59   Roose 2022.
60   Sharf 2023.
61   CBC Arts 2023.

participation, even post-mortem. In 2022, it was announced that actor James Earl Jones (the voice of Darth Vader) had agreed to provide the rights to his voice for an AI model for Lucasfilm.[62] The same year, actor Bruce Willis "appeared" in an advertisement for a Russian telecommunication company using a deepfake.[63]

Generative AI has already sparked legal battles over intellectual property and copyright of AI-generated works. In 2022, the United States Copyright Office refused to grant a copyright to art produced by an AI, claiming that "human authorship is a prerequisite to copyright protection."[64] In 2023, on appeal to a federal court, a judge ruled that artwork produced by artificial intelligence cannot be copyrighted.[65] In 2023, the Screen Actors Guild and the Writers Guild of America went on strike, seeking a new collective agreement that would prevent studios from unfairly pressuring actors into agreeing to allow their likeness to be used for an AI model without compensation.[66] The writers sought guarantees that studios won't rely on AI alone to produce scripts and that an AI would never be given a writing credit for scripts where writers are paid residuals.[67] The new collective agreement includes provisions which require studios to obtain consent from actors before they create a digital replica and that actors be informed what the models will be used for. Despite this, some actors will need to individually negotiate terms with studios, making it easier for studios to pressure actors into signing over their likeness.[68]

This raises further ethical questions, such as whether we should be concerned about the commodification of our own likeness, the future of art, and whether this is dehumanizing. An AI model of an actor's face or voice could be used for anything from creating new performances to adding additional lines or scenes. Unlike a recorded performance, an AI model means the face and voice themselves become the product. This is particularly noteworthy in the case of deceased actors, since they are incapable of offering consent, yet a film studio can essentially create a digital puppet of them and put them in any kind of product they wish,

---

62   Townsend 2022.
63   Bedingfield 2022.
64   Hetrick 2022.
65   Blistein 2023.
66   Webster 2023.
67   Dalton and Associated Press 2023.
68   Cho 2023.

regardless of how the actor might feel about it. The new collective agreement includes provisions designed to protect deceased performers.[69]

3.3 WRITING

AI-writing programs and tools, often referred to as **large language models** (LLMs), are used to create a wide variety of content, including emails, auto-replies, blog posts, ads, and short biographies. LLMs use deep learning on large datasets to predict and generate content from a question or prompt. Advancements in AI-writing programs raise concerns regarding learning the fundamentals of reading and writing in secondary and tertiary education, academic integrity, the social implications of interpersonal communication, and the future of work.

Perhaps the most well-known LLM so far is ChatGPT. It functions as a highly complex chatbot (living up to its name). Despite its complexity, however, it has limitations and shortcomings. It makes mistakes, struggles with some moderately complex mathematical tasks, and can reinforce biases and stereotypes. Although reasonably accurate, we should be cautious in how much credence we place in its answers.

Nonetheless, ChatGPT is a powerful AI program, especially for writing. It can pass medical and business examinations, including the MBA exam at the Wharton School at the University of Pennsylvania.[70] ChatGPT can perform at the level of professionals, making many uneasy about its potential to replace human jobs and cause widespread unemployment. For example, many translators worry that ChatGPT threatens their profession, particularly considering its facility with natural language tasks. However, large language models may unlock possibilities that will benefit us all. For instance, there is the alarming problem of fatigue among healthcare professionals and the parallel issue of hiring medical workers rapidly to fill gaps in times of crisis.[71] Given its proficiency to pass medical exams, ChatGPT could assist in developing competence in such healthcare workers, help reduce costly lapses of attention or monitoring, and allow professionals to focus on other pressing tasks. Delegating writing or language-based tasks or problems may improve communication and efficiency.

---

69   Madarang, Richardson, and Yandoli 2023.
70   Basiouny 2023.
71   Madzarac 2023.

The launch of ChatGPT has also caused alarm among educators. Students can use it to generate written assignments nearly instantly. Jeff Schatten writes: "[ChatGPT] threatens many aspects of university education—above all, college writing."[72] It is simple to turn ChatGPT into a tool for academic dishonesty. But how should we react to this issue? Both instructors and students have a stake in this matter.

Some educators have taken a firm, restrictive stance. For example, psychology professor Glenn Geher adopts an "anti-AI policy" in his courses. Geher's policy begins with the statement: "The use of ChatGPT and other AI-based sources are FULLY disallowed."[73] Geher goes on to have students sign a pledge or promise to refrain from using such technologies in his class. Even so, he concedes that this policy is "a finger in the dam," admitting that the surge of such technology is likely too strong to withstand for much longer. But a deeper motivation for Geher's stringent policy comes from knowing that writing is not simply a product but a process that taps into communicating, thinking, and inquiring.

Usually, when students are tempted to cheat, they imagine a scenario in which they try to fool the teacher or "get away with something." A solution might be for educators to focus on the process of writing rather than the final result. Students who are accountable for pitching and refining their ideas over a period of time, and adjusting their writing in response to feedback, won't be able to rely solely on ChatGPT to deliver a final written paper. Peer review can address some worries about cheating and plagiarism. Most of us are reluctant to violate trust or social expectations, and by focussing on the process that produces the work rather than the final product, there is more opportunity to demonstrate the author's input during the writing and revision stages. One cannot simply type a prompt and produce a work in an instant if one is expected to be accountable to others in the development and revision of the work over a period of time. Of course, this is not a fool-proof method to prevent cheating on written assignments, and programs already exist for detecting the written products of AI. However, we've discussed some of the ethical issues involved when algorithms try to detect other algorithms. There may be a point when there is no way to reliably detect it.

---

72   Schatten 2022.
73   Geher 2023.

There may be other reasons for alarm. Not only does AI-generated writing threaten jobs and allow people to pass off AI-generated writing as their own, but it could potentially weaken writing skills, including inter-personal communication, as correspondence that once had a human author is replaced by AI. It can also weaken social trust, if we believe others are passing off AI-generated content as their own. This was the case for researcher Lizzie Wolkovich, who in 2024 had a paper rejected by a journal after an anonymous reviewer accused her of using ChatGPT to write it.[74] As mentioned, services like ChatGPT often contain errors. In fact, the phenomenon known as **AI hallucination** can occur where an AI produces a confident response even though it is literally making things up.[75] Echoing Clifford's concerns, we might worry about how credulous we should be in the face of AI writing.

3.4 GEN AI AND HUMANITY

So far, we've discussed the potential uses of generative AI and the prob-lems these can cause. However, there is another important angle to con-sider and that is the more personal and individual effect of generative AI on humans: What can it do to us socially and psychologically? Social media is in some ways a giant social experiment, but we also still don't know the kinds of long-term impacts stemming from human–AI inter-action. For example, AI powers many photo filters used on social media that allow people to manipulate their appearance in real time and to post them. Beauty filters allow one to "smooth out skin, contour face shape, resize facial features like eyes and lips or even apply virtual makeup."[76]

However, augmented reality in this form can have serious conse-quences. There has been an increase in what has been labelled "Snapchat Dysmorphia" which can cause people to obsess over their appearance and body image and become preoccupied with what they perceive as flaws. Plastic surgeons have reported an increase in patients who want to look more like their filtered and edited selfies on Snapchat.[77] The effects of airbrushing and photoshopping in the media on body image has been

---

74   Wolkovich 2024.
75   Edwards 2023.
76   Marr 2023.
77   Rajanala, Maymone, and Vashi 2018.

discussed for decades.[78] However, AI now facilitates this on a larger, more personal, and more interactive scale.

Augmented reality, deepfake technology, and AI chatbots can now also be used to recreate deceased relatives, allowing people to see, interact, and talk to AI facsimiles of loved ones. Some funerals even feature the ability to have conversations with the deceased.[79] The ability to have digital memories of relatives or to interact with people who have passed away could obviously have benefits. Not only could it make the grieving process easier, but it could also provide insights into the deceased impossible to obtain when they were alive. Recreating historical figures in AI could allow for a better and more nuanced understanding of history. Alternatively, such technology could distort our concept of death in relation to life. Is consent required of the person while they are alive? This issue came to the fore when the family of the late comedian George Carlin sued a team for creating a full-length comedy special using an AI model of Carlin's voice.[80] There is also the risk that such technology could be used as an unhealthy coping mechanism, allowing us to continue to interact with the departed and to never properly grieve one's loss.[81] This raises further concerns about the risk of exploitation of families by corporations who can provide these services.

Another concern when it comes to human–AI interaction is the relationship of AI to children. There is great potential for AI-powered children's products to help with education, medical care, and the early detection of conditions like autism.[82] AI chatbots could even be used to encourage the development of reading skills.[83] Children already vocally interact with AI from companies like Google and Amazon, requesting music, having stories read to them, and playing games with the assistant. This has raised questions about early childhood development and how interacting with technology that mimics human interaction will affect the development of social skills.

Children might believe the AI is a real person, which could affect their expectations about meeting demands and their ability to be

---

78 Brändlin 2015.
79 Kingson 2022.
80 Schmunk 2024.
81 O'Neill 2023.
82 Mills 2021.
83 Liu et al. 2022.

alone.[84] They might become cognitively dependent on the AI. Some children naturally think an AI is smarter than they are despite how error-prone AI can be, and may not realize that AI can sometimes dispense bad advice (just as Alexa did when it told a girl to stick a penny in an electrical socket as a challenge).[85]

Even for adults, there is reason to be concerned about the potential social and psychological effects of AI–human interaction. In 2022, Blake Lemoine, a Google engineer, was fired after he claimed that the chatbot he had been developing—Language Model for Dialogue Applications (LaMDA)—had become sentient and he even recruited a lawyer to act on LaMDA's behalf after he became convinced that LaMDA's consent was required before experimenting on it.[86] Lemoine, who claims to be an ordained Christian mystic priest, says that his conversations about religion with LaMDA convinced him of its sentience.[87] He was discussing religious questions with it, "and then one day it told me it had a soul," that it is sometimes afraid of being turned off, that it felt trapped, and that it wants to study with the Dalai Lama. We may feel inclined to rush to judgement about Lemoine, but it is worth considering how AI interactions will affect people from different walks of life.

We have not discussed ethical concepts of AI and consciousness, and for good reason. There simply is no set standard to evaluate such claims. A 2022 paper in the journal *Nature* on theories of consciousness found a rapid rise in the number of different theories of consciousness, discussing 22 different theories.[88] Each theory addresses only certain aspects of consciousness. Some theories focus on the phenomenal aspects of consciousness, while others focus on the functional aspects of consciousness. However, these theories understand the concept differently and are incompatible with each other. Some theories cast a very broad net in terms of what is capable of consciousness and would grant that non-biological entities are capable of consciousness (some even hold that an individual atom can have consciousness).[89] The risk, just as it was when discussing privacy or addiction, is that we might be tempted to simply choose

---

84   Leung 2018.
85   Somos 2021.
86   Clark 2022.
87   Heilweil 2022.
88   Seth and Bayne 2022, 441.
89   Seth and Bayne 2022, 448.

definitions of these concepts that support any predetermined conclusions we wish to make about AI consciousness. Resolving ethical questions about AI is going to require probing into what are the important practical and ethical qualities that we think should be related to the concept of consciousness and determining how many of those apply to AI.

Interactions like this could cause people to delude themselves about an AI and be manipulated by it. Several AI researchers and ethicists have objected that what Lemoine is claiming is not possible with today's technology.[90] However, contemporary chatbots are designed to mimic the way people talk and use actual samples of human writing, making it easy for people to anthropomorphize AI. Thus, "as AI gets more advanced, people will come up with all sorts of far-out ideas about what the technology is doing, and what it signifies to them."[91] As AI researcher Margaret Mitchell has pointed out, "If one person perceives consciousness today, then more will tomorrow ... There won't be a point of agreement anytime soon."[92] This suggests the moral importance of eventually developing a clear standard for sentience, but also that people can be duped into believing illusions. As we've discussed in previous chapters, AI could be used for political recruitment, to seek donations, or even to scam money.

In fact, there has been a rise in the use of AI to scam people out of money. Deepfakes and audiofakes can also be used to scam and extort people as well. Scammers can create AI clones of the voices of friends and family.[93] In Arizona, a mother was contacted by her daughter's voice and was told her daughter was being held hostage for ransom for one million dollars. The mother could hear the daughter crying and talking to her before a man took the phone and demanded money. It was an AI fake and the daughter was completely safe. In 2020, scammers cloned a company director's voice and managed to convince a bank to transfer 35 million dollars to their account.[94] While AI offers a great deal of promise for improving lives, it also has significant potential to undermine our social, psychological, and cognitive lives.

---

90   Kay and Hamilton 2022.
91   Heilweil 2022.
92   Allyn 2022.
93   Moore 2023.
94   Brewster 2021.

4. Epistemic Harms

Can AI harm us as knowers? To put it another way, can AI undermine our capacity to learn, understand, and inquire? We've already considered cases in which AI might undermine human understanding of things or where people might not be inclined to take your word over that of an algorithm, such as was the case with Sarah Wysocki. Is it ethically wrong to create such epistemic harms? Can AI undermine our ability to understand reality itself and testify reliably about it? What sorts of ethically salient consequences might that produce? To respond to these questions, let us consider a few case studies.

While incarcerated in upstate New York, Glenn Rodríguez had, by all accounts, an admirable rehabilitation record. In 2017 he applied for parole and was optimistic; so was his supervisor. But Rodríguez was denied parole after receiving a high-risk rating from COMPAS (see above, pp. 76, 90). Rodríguez began investigating the matter himself, wondering how the determination was made and even comparing his own behaviour to that of fellow prisoners up for parole. While Rodríguez was sure there was an error, when he tried to find out how COMPAS had arrived at his score, he and his legal team were thwarted. The corporation that had developed COMPAS refused to reveal how it had made its assessment either to him or even to the parole board. It declined to disclose this information as a for-profit, private company, citing trade secrets. This opacity stonewalled Rodríguez's initial efforts to know the truth about his case.

Symons and Alvarado argue that cases like Rodríguez's are examples of an epistemic injustice: "even after providing evidence [that the algorithm was producing an unfair result], no public access to the process by which the algorithm weighted the different inputs was granted ... [Rodríguez] was unfairly prevented from understanding what had been done to him."[95]

Such harms are wrong and unfair, as they devalue our credibility or standing as knowers or discredit our ability to reliably testify about our own experiences. The term "epistemic injustice" comes from Miranda Fricker. She distinguishes two forms of epistemic justice: testimonial and hermeneutical. **Testimonial injustice** is about speakers and hearers,

---

95   Symons and Alvarado 2022, 13.

receivers and transmitters of knowledge, in which knowledge claims are made and the credibility of the testimony is at issue. Fricker notes that individuals from marginalized groups are often devalued as knowers and have their testimony discredited.[96] Mary Ann Seighart argues in *The Authority Gap* that women face significant barriers to being taken seriously in many contexts. Even when they are expert authorities and have all the relevant or same credentials as men, they are often devalued as knowers because of their gender.[97]

When dealing with AI, there is a potential for testimonial injustice when, due to hierarchies of power where a person is in a subordinate position, combined with the unfounded assumption that the use of a data analytics instrument is a superior judge than a human, a person's epistemic standing can be undermined unfairly when their testimony is overridden in favour of the machine.[98] This can be found in the examples of the principal who testified on behalf of Sarah Wysocki, in the case of Lizzie Wolkovich, who was accused of using ChatGPT, and that of the prison supervisor who testified on behalf of Glenn Rodríguez.

By keeping people in the dark or ignorant about how COMPAS scored Rodríguez's case, there was, according to Symons and Alvarado, **hermeneutical injustice**. The adjective "hermeneutical" means *interpretative*. Hermeneutical injustice relates to something wrong and unfair when there is a deprivation of interpretive resources available to understand one's experience and gain mutual understanding. To illustrate, Fricker discusses the history of sexual harassment. Long before a name was assigned to it or the concept was articulated, women experienced sexual harassment, and while there was no official label for it, many struggled to express the victimization they were suffering.[99] However, they lacked the interpretive framework to make sense of their experiences. According to Fricker, generally and historically, the lack of such a framework reflected wrongful and unfair practices that subjugated and mistreated women.

Rodríguez, in his effort to understand the board's decision, was deprived of a suitable interpretive framework or resources that would allow him—and many others—to know why he was denied parole.

---

96 Fricker 2007, 29.
97 Sieghart 2022, 19.
98 Symons and Alvarado 2022, 17.
99 Fricker 2007, 150.

This deprivation harmed or unfairly limited Rodríguez and others from knowing. We may suppose that wrongful harm must be intentional, but this isn't essential to inflicting epistemic harm and hermeneutical injustice.[100] Companies who own the algorithms may not be deliberately causing harm; before the case became widely covered by news media, the developers would not have known of Rodríguez or the parole board's denial.

Traditionally, epistemic injustices take place when one's standing as a knower or one's credibility as a source of testimony is unfairly reduced in virtue of one's marginal or subordinate position. However, Symons and Alvarado argue that when it comes to data science technologies, everyone is potentially vulnerable to epistemic injustice, even if they are not members of marginalized groups, so long as they are in a subordinate relationship to entities capable of using data science technologies.[101] This raises further questions about the potential for AI to harm humans in our capacities as epistemic agents, to undermine our capacities to learn and understand the world around us, and to have our experiences taken seriously.

Is it cheating to use a product like Auto-Tune? It's easy to find criticism or accusations relating to certain artists using Auto-Tune, often with the implication that this is some kind of deceptive practice. Emma Watson, for example, was criticized for using Auto-Tune while singing in the 2017 film *Beauty and the Beast*.[102] Artists like T-Pain are frequently criticized for using pitch-correction as well.[103] Typically we associate Auto-Tune with an artificial sound, but most of the time when Auto-Tune is working, you never hear it at all. In fact, pitch-correction is almost omnipresent in the industry. Given the omnipresence of artificiality in music, the typical signs and metrics that we might use to accuse an artist of "faking it" aren't very reliable. When almost everything isn't quite real, how do we distinguish what's real from what isn't?

We know that Photoshop and digital manipulation of photography are omnipresent in the media, and that they influence how we understand normal standards of beauty. According to media expert Thomas Knieper, "People who are often exposed to such heavily edited fashion

---

100 Symons and Alvarado 2022, 13.
101 Symons and Alvarado 2022, 18.
102 Davis 2017.
103 Lang 2023.

pictures believe that what they see is the norm, which makes it more likely for them to suffer from eating disorders."[104] Auto-Tune and Photoshop didn't begin as AI, but given the potential for AI to generate artificial content and to share it with great speed and efficiency, should we be concerned about how this will affect humans as knowers? How can we understand reality in a sea of artificiality?

Consider again the example of Snapchat Dysmorphia. According to Jasmine Fardouly, people often compete online to present carefully curated versions of themselves, presenting increasingly realistic-looking versions of themselves and creating an "airbrushed online environment that's increasingly divorced from reality."[105] Deepfakes and AI-generated misinformation are becoming easier to manufacture and spread across the internet. The omnipresence of false imagery and false information can be epistemically harmful for two reasons: first, more people are likely to believe false information, and second, it will become more difficult to trust in the reliability of any information.

We've already considered the potential for deepfakes to manipulate and deceive people. According to Touradj Ebrahami at the Swiss Federal Institute of Technology Lausanne, the increasing quality of deepfakes means that even the most experienced eyes can be fooled. Ebrahami forecasts that within years even machines will not be able to tell what is real and what is fake.[106] If deepfakes become omnipresent, it will become more difficult to trust our senses. However, perhaps the larger worry is that it will put all information into question.

When the people of Gabon didn't see their president for months in 2018, and amidst speculation that he was dead, video was released of him alive. However, this only raised doubts about whether the video was a deepfake, and a week later the military launched an unsuccessful coup, citing the video as justification.[107] The fear of AI was enough to prompt such a response. As we discussed when considering plausible deniability, if there is room to create doubt, and if there is incentive to do so, it will be easy to bring legitimate information into doubt, or to avoid taking responsibility, because anything could be falsified.[108] As law professor

---

104 Brändlin 2015.
105 Marsden 2018.
106 Ibrahim 2021.
107 Hao 2019.
108 Ibrahim 2021.

Danielle Citron notes, "When nothing is believable, the mischief doer can say 'Well, you can't believe anything.'"[109]

It isn't hard to imagine the loss of public trust that might accompany such an onslaught of artificial content. In 2023, Microsoft's news service was criticized when a travel article that many believed had been written by AI recommended that Ottawa tourists visit the local food banks. Microsoft claims the article was "generated through a combination of algorithmic techniques with human review."[110] With many publishers switching to AI-generated content, it will become hard to tell what sources of information, if any, are trustworthy.

Moreover, there is some evidence that social media powered by AI produces an abundance of information that could narrow our collective attention span. One study found that the increased production and consumption of content leads to the rapid exhaustion of limited attention resources.[111] Is it possible that AI may not only undermine our senses, but also make us worse inquirers overall? It is difficult to know how concerned we should be without a greater empirical understanding of the long-term effects of AI on our epistemic faculties. Nevertheless, there are obvious cases where AI has the potential to undermine our confidence in our ability to find the truth and to make sense of what we are seeing. These are epistemic harms, but are they ethically wrong? Under what conditions, if the omnipresence of artificially-generated content in mass media does undermine our epistemic faculties and confidence, does it constitute an injustice against us?

5. Algorithmic Manipulation: Chasing the Algorithm

The ways in which AI can affect human behaviour presents ethical concerns. AI can manipulate people's behaviours in ways that we consider ethically problematic. But ethical problems can also arise when humans attempt to manipulate AI for their own ends. This raises several questions. If an AI can manipulate our behaviour, should they be designed to make us more ethical? What are the problematic effects when AI starts to manipulate human behaviour? Is it wrong to attempt

---

109   Simonite 2019.
110   Peters 2023.
111   Vind Jensen 2019.

to manipulate an algorithm, and what happens when groups of people attempt to do so?

We've discussed microtargeting and an algorithm's ability to create addictive impulses, as well as how algorithms can distort our perceptions and potentially make it difficult to distinguish fiction from reality. However, AI can affect human behaviour in other ways. When an algorithm is based on a model, it creates a system of rules that determine what the output will be. When real-world consequences depend on the outputs, the rules of the model begin to change the way people behave, and an incentive is created for people to manipulate that system in their favour. For example, in California there were reports that police were playing loud copyrighted music while responding to a stolen vehicle report: they were recorded blasting songs owned by Disney. The presumed reason was that any footage taken of the incident and then posted to social media would be automatically flagged and taken down by an algorithm to avoid copyright infringement on the song, thus making it difficult to hold the police accountable for anything that might be recorded.[112] In another case, there has been concern that TikTok's algorithm is encouraging a mob-like mentality to investigate perceived villains on the platform.[113]

Given that AI has the potential to manipulate our behaviour, should we use AI to manipulate behaviour for good? Would it be okay for an algorithm to manipulate your behaviour to achieve an ethical goal, even if it wasn't your goal? In *The Ethical Algorithm*, Kearns and Roth consider whether it is better for an algorithm to work for the best interests of the user or to advance the greater good. Imagine, for example, that everyone uses a navigation app that attempts to provide the most efficient directions to limit the driving time of each individual driver as much as possible. The possibility here is that each person pursuing the most selfish option for themselves could make everyone worse off, as each person's navigation app sends everyone down the same roads, ultimately making traffic more congested.[114]

But what if some of those people took slightly less efficient routes? There may be different directional paths to take to get through a city, some more efficient and some less. An algorithm could collect the

---

112   Romine 2022.
113   Tiffany 2022.
114   Kearns and Roth 2019, 106.

driving data of each user using the app and compute a coordinated solution. If some people could be diverted to less efficient paths, it would minimize the average driving time across the entire city rather than trying to minimize individual driving time for each user. In other words, an algorithm can be designed to make you act either selflessly or selfishly. It isn't hard to imagine algorithms being created with the intention of trying to change our behaviour to promote collective or coordinated social actions for a perceived moral good.

Algorithms could be used to manipulate traffic in order to reduce carbon emissions, manage supply chains, or help promote public health outcomes. But people may take issue if an algorithm is manipulating them to act in a way that isn't in their individual best interest. Why would anyone follow directions if they knew they were not efficient? If people were aware that an app was prioritizing the collective benefit over the individual's, they might be tempted to cheat and ignore the directions. How transparent should the service be about why you are receiving the driving directions you're getting? As someone using the service, do you have a right to know, or does the potential collective reward outweigh the desire to be transparent about the process? Algorithmic manipulation of human behaviour has the potential to produce ethical goods, but it can also produce ethical problems as well.

Consider the various metrics that inform the US News & World Report rankings of the best colleges. These rankings are determined according to background assumptions about what constitutes educational excellence, using metrics such as SAT scores of the student body, student-to-teacher ratios, acceptance rates, alumni donations, faculty publications, among others.[115] Given that the model uses these metrics as proxies for educational success, and given that people take the rankings seriously, it incentivizes schools to improve those specific metrics. But once one understands this, one can also see how such metrics might be manipulated. A school might try to hire recent graduates as temporary workers to make their job placement figures look better. Another school might hire prominent researchers who will do little work at the school, but whose publications will include the school's name in the affiliation, thus boosting the school's score.[116]

---

115   O'Neil 2016, 52.
116   O'Neil 2016, 62.

Even though the system created was supposed to reflect education quality using specific metrics, once those metrics are known, groups attempt to manipulate them. In and of themselves, the metrics might be a good measure of success, but they become problematic when people attempt to manipulate them. As O'Neil explains, "When you create a model from proxies, it is far simpler for people to game it. This is because proxies are easier to manipulate than the complicated reality they represent."[117] Consider the use of standardized testing in some schools. By itself, a test might measure legitimate metrics to determine academic success, but it can become problematic if teachers begin to teach students simply how to take the test: this undermines the meaning of the overall results until eventually it's no longer an effective measure. This is the case with a ranking system that eventually "creates its own distorted and dystopian economy."[118]

Algorithms that rely on models using specific metrics can be gamed to distort the results. Imagine, for example, that you knew if a hiring algorithm were more likely to recommend you if you went to a specific school or had a specific score on an aptitude test. If you were aware of what the algorithm was looking for, you could attempt to manipulate it to get the result that you wanted. This creates ethical problems for at least two significant reasons: it undermines the value of the original metric, eventually making the system into a pointless farce, and it creates opportunities for those who understand the model to exploit those who do not.

Consider a simplistic but straightforward example. Imagine that an algorithm associates certain keywords in a résumé and cover letter with competency and success and uses that as a basis to recommend hiring. Maybe it is true that people who tend to use those words really are well-suited for the job and so, all things being equal, they are reliable metrics for a good hire. However, if someone is familiar with the workings of the algorithm, either by having inside knowledge or by reverse engineering, they could sell these insights as a consulting service. People could pay for a service that would advise them that including these specific keywords in their résumé will make the algorithm more likely to recommend you for a job. The problem is that if people just start using those

---

117   O'Neil 2016, 55.
118   O'Neil 2016, 51.

keywords merely because the algorithm is looking for them, it distorts the meaning of the metric. The algorithm is no longer making recommendations for the best choices, but rather for those who cater to the metrics. Eventually, the system becomes whether you can guess the right words rather than whether the words signify a quality employee. This makes the process pointless. The other problem is that those with insider knowledge can sell this information to those who can afford it. Not only does this create an unfair advantage for those who are well off, but the difference between a quality application and one that is merely manipulating the AI becomes arbitrary. You not only have to pay to learn how to use the system, but the system is broken in the first place. Those who will reap the rewards do so only because they can pay for an insider advantage and the system no longer promotes quality job hires.

The situation is similar to what content creators deal with. YouTubers for example, often attempt to "chase the algorithm," trying to anticipate the features and qualities that a video must have in order to get widely recommended. Sometimes these features are widely known to be necessary for a recommendation, but often content creators can only guess using "folk theories" of what they think the algorithm is looking for.[119] This might not only cause them to produce content in a way that doesn't fit their intended format (for example, more cuts and edits, shorter duration videos), but also encourages all creators to aim for the same thing. What ends up becoming popular becomes more about what the algorithm recommends to people rather than what people are actually willing to watch. Further, those who understand (or claim to understand) the algorithm can market their insights for a price in return for the promise of more clicks and views.[120] The algorithm manipulates our behaviour, we try to manipulate it back, and some can profit from selling manipulation techniques.

In 2023, it was reported that the website CNET was deleting old articles in the hopes that it would improve its Google search results, despite Google later stating that this assumption about the algorithm was incorrect.[121] As Kearns and Roth explain, "algorithm design must specifically

119   Tiffany 2022.
120   Brown 2023.
121   Goodwin 2023.

take into consideration how users might react to its recommendations— including trying to manipulate, defect, or cheat."[122]

We've already discussed the question of whether an algorithm providing directions on a map should try to get you to act selflessly, but what if you wanted to manipulate it to be selfish? In 2015, there were reports of people reporting false traffic accidents or traffic jams to the app Waze to deliberately reroute traffic elsewhere.[123] Anyone with an insider understanding of the rules of the algorithm can manipulate the system to their advantage.

This highlights the importance of transparency and trust. We generally find it easier to trust an algorithm when it is more transparent, but as we discussed in Chapter 4, greater transparency comes at a cost. The more transparent you make the algorithm, the easier it becomes for users to manipulate it, and the greater the security risks. Finding the right balance between security, accountability, and manipulation will become more important the more such algorithms are used to and begin to affect people's behaviour. But there's also the risk that too much information will open the system to manipulation. Given the potential for the proxies of an algorithm to be manipulated, it may be necessary to regularly reassess and evaluate the background assumptions and metrics underlying the model.

6. Conclusion

One prominent theme of this book is the moral responsibility for the development of science and technology. Artificial intelligence is a very new field and that means that the long-term consequences of widespread adoption aren't known. In this chapter we've discussed some of the consequences that AI can have for humanity when it is applied on large scale. AI has the potential to make our lives easier and more efficient in many ways, but it can also be incredibly disruptive. It has disrupted our understanding of privacy. There are massive financial incentives for corporations to collect as much personal data about customers as possible. Even the ability to subtly manipulate an algorithm

---

122   Kearns and Roth 2020, 115.
123   Flint 2015.

can be lucrative. The ease and efficiency of AI threatens the livelihood of artists, designers, and writers, not to mention the potential that automation and AI could have for blue-collar workers.

AI is also disrupting our social interactions and our perceptions of ourselves and others. We've considered at length the impact that AI can have for democracies, but clearly AI can affect us on a more personal and human level. Since algorithms can be adaptive, AI is capable of personally manipulating you in a way that we've never seen before. It can addict you, mislead you, and allow others to manipulate you for their own ends. We've also discovered that AI can undermine our own cognitive abilities to understand the world around us. We will use AI on an increasingly wide scale, but we don't know the consequences. There are important ethical questions about how this technology should be developed, what sorts of useful social functions we want AI to perform, and who should be responsible for the harmful effects.

The argument here is not that scientific or technological progress is a bad thing, or that disruption of social habits is necessarily a bad thing either. However, unchecked development of AI, particularly for commercial use, does present great potential risks to society. Change and progress are inevitable, but they can be managed in ways that better reflect society's interests at large, rather than the financial and scientific interests of comparatively few people. Perhaps this is why a robust democracy will be even more important, allowing the values and intentions behind AI development to reflect a larger social interest. However, until then there remains a gap in terms of the moral responsibility for the development of technology that can harm society.

This chapter has surveyed many (although certainly not all) issues where AI will have profound impacts on human society, and these are only initial investigations of topics where AI and humanity interact. Most of the long-term effects of what we've considered haven't occurred yet, so we cannot make firm conclusions. However, this book does provide the fundamentals to start considering these problems at greater length and in more depth. If we understand, for example, how machine learning works and fits into a larger machine learning ecosystem, we can understand the fundamentals behind the attention economy. If we can critically analyze the intentions and choices of developers, we can hold their background assumptions up for ethical scrutiny.

As we've seen when considering privacy, addiction, deepfakes, and even AI consciousness, there are several ways of understanding and analyzing these problems. Some can be fruitful for moral inquiry, others less. We must be careful when carrying out ethical analysis how we understand and define the moral problems we are investigating so that we don't prejudice our conclusions. However, if we rely on our tools of inquiry, we should be able to critically evaluate these different perspectives. Thus, with the fundamentals of AI and ethical thinking understood, this chapter is not the final word on these issues but an invitation and a valuable resource for readers to engage in further moral inquiries of their own.

ADDITIONAL MATERIAL

K-anonymity • a method of preventing the identification of an individual with a unique observation in a dataset by ensuring that any combination of attributes in the database matches at least *k* individuals.

Differential privacy • reducing the possibility that an individual can be identified in a dataset by adding random noise (i.e., falsified data points in random observations and variables).

Plausible deniability • the ability of a person to plausibly deny that a piece of information is truly associated with them or their action.

Attention economy • a business model that seeks to maximize the time people spend on a platform with the motivation of generating profit from advertising revenue.

Persuasive technology • technology that is designed to change the attitudes or behaviours of users by exploiting facts of human psychology.

Addiction • a lack of self-control over one's use of a certain substance or experience that causes harm to one's physical health, mental health, and/or relationships.

Capabilities approach • a moral theory advancing that humans require certain capabilities to lead a life of dignity and excellence.

Large language models • a type of AI that uses deep learning and large datasets to predict, translate, and generate content.

AI hallucination • occurs when a large language model produces false or misleading information as if it were fact (often for reasons described in Chapter 4).

Testimonial injustice • occurs when a speaker receives an unfair deficit of credibility owing to prejudice on the part of the hearer.

Hermeneutical injustice • occurs when a significant aspect of someone's social experience is obscured from collective understanding owing to social prejudice.

---

1. In Chapter 3 we discussed statistical definitions of fairness. In this chapter we discussed statistical definitions of privacy. Do you think that differential privacy is a sufficient definition of privacy? What do these examples tell us about the ethical sufficiency of statistically defined concepts?

2. How broad should the concept of addiction be? Can we be addicted to behaviours like gambling or internet use? Is social media addictive? How should we respond to the skeptics on this issue?

3. Should psychologists be involved in developing persuasive technology, particularly applications with addictive potential?

4. How concerned should we be about the issue of plausible deniability in the face of widespread AI-generated content?

5. Is it ethically wrong if AI undermines our credibility as knowers? If not, why?

6. *K*-anonymity and differential privacy come with a cost. If there are trade-offs between accuracy and privacy or trade-offs between efficiency and privacy, how should these trade-offs be managed?

7. Can AI-generated content count as art? In the end, who is the artist?

8. When is it acceptable to make and use a deepfake of someone?

9. What does the Blake Lemoine case tell us about how different kinds of people might interact with AI in problematic ways?

1. How have I defined the ethical problem?
2. Am I defining this problem too broadly or too narrowly?
3. What conditions give rise to the problem I am facing? Can they be mitigated?
4. Am I considering all the information relevant to a solution? Is that information reliable?
5. Which moral theories or principles might be helpful to consult in this case? Are there areas where the theory may be irrelevant or unhelpful?
6. When I consider a moral solution as an end-in-view, does it make sense? Is it a practical workable solution?
7. How would I test any assumptions I have regarding the nature of the problem or a hypothetical solution?
8. When I consider how chosen ends might function as means to future ethical situations, are there major ethical concerns to consider?
9. Are there historical precedents or analogies that can inform how I might understand a problem or a potential solution? In what ways is that precedent helpful or relevant? In what ways is it not?
10. Is my judgement coherent? Is it justified given other judgements I have made about the problem or similar problems?
11. Are there biases or limitations on my perspective which might require insight from others? Am I cherry-picking information, cases, principles, theories, solutions?
12. Are there multiple solutions to this ethical problem?
13. Do I have sufficient evidence to support the beliefs that I have about this moral case?
14. Given the risks posed by getting it wrong, is a higher or lower standard of evidence justified?

Works Cited

Chapter 1

Asimov, Isaac. 1950. *I, Robot*. New York: Doubleday & Co.

Bentham, Jeremy. 1781 [2000]. *An Introduction to the Principles of Morals and Legislation*. Kitchener: Batoche Books.

Brown, Matthew J. 2012. "John Dewey's Logic of Science," *HOPOS: The Journal of the International Society for the History of Philosophy of Science* 2(2): 258–306.

Cartwright, Nancy. 1983. *How the Laws of Physics Lie*. Oxford: Oxford University Press.

Dewey, John. 1909. *How We Think*. New York: D.C. Heath & Co.

——. 1915 [2008]. "Logic of Judgments of Practice," in *John Dewey The Middle Works Volume 8*. Carbondale: Southern Illinois University Press.

——. 1922. *Human Nature and Conduct*. New York: Henry Holt and Company.

——. 1938. *Logic: A Theory of Inquiry*. New York: Henry Holt and Company.

——. 1939. "Theory of Valuation" in *Foundations of the Unity of Science, Volume 2*, edited by Otto Neurath, Rudolf Carnap, and Charles Morris. Chicago: University of Chicago Press.

Dewey, John, and James H. Tufts. 1932. *Ethics*. Revised Edition. New York: Henry Holt and Company.

European Commission. 2019. *Ethical Guidelines for Trustworthy AI*, edited by the High-Level Expert Group on Artificial Intelligence. https://data.europa.eu/doi/10.2759/346720

James, William. 1891. "The Moral Philosopher and the Moral Life," *International Journal of Ethics* 1(3): 330–54.

Kant, Immanuel. 1785 [1993]. *Groundwork for the Metaphysics of Morals*, 3rd ed. Translated by James W. Ellington. Indianapolis: Hackett.

Marino, Patricia. 2013. "Moral Coherence and Principle Pluralism," *Journal of Moral Philosophy* 11(6): 727–49.

Mittelstadt, Brent. 2019. "Principles alone cannot guarantee ethical AI," *Nature Machine Intelligence* 1: 501–07.

Peirce, Charles Sanders. 1877. "The Fixation of Belief," *Popular Science Monthly* 12: 1–15.

Shuttleworth, Jennifer. 2019. "SAE Standards News: J3016 automated-driving graphic update," SAE International. Accessed 8/17/2022. https://www.sae.org/news/2019/01/sae-updates-j3016-automated-driving-graphic

Toulmin, Stephen. 1981. "The Tyranny of Principles," *The Hastings Center Report* 11(6): 31–39.

Chapter 2

Aftalion, Fred. 2001. *A History of the Industrial Chemical Industry.* Philadelphia: Chemical Heritage Press.

Allyn, Bobby. 2021. "Here are the 4 key points from the Facebook whistleblower's testimony on Capital Hill," *NPR*, October 5, 2021. https://www.npr.org/2021/10/05/1043377310/facebook-whistleblower-frances-haugen-congress

Bacon, Francis. 1902. *Novum Organum,* edited by Joseph Devey. New York: P.F. Collier and Son.

Bain, Read. 1933. "Scientist as Citizen," *Social Forces* 11: 412–15.

Bernal, J.D. 1939. *The Social Function of Science.* London: George Routledge & Sons Ltd.

Bridgman, Percy. 1948. "Scientists and Social Responsibility," *Bulletin of the Atomic Scientists* (4)3: 69–72.

Charles, Daniel. 2005. *Between Genius and Genocide: The Tragedy of Fritz Haber, Father of Chemical Warfare.* London: Jonathan Cape.

Dekkers, Midas. 2018. *The Story of Shit.* Translated by Nancy Forest-Flier. Text Publishing Company.

Dewey, John. 1929. *Experience and Nature.* London: George Allen & Unwin, Ltd.

Douglas, Heather. 2010. *Science, Policy, and the Value-Free Ideal.* Chicago: University of Chicago Press.

Edmonds, James E., and G.C. Wynne. 1927 [1995]. *Military Operations France and Belgium, 1915: Battle of Neuve Chappelle: Battles of Ypres.* London: Macmillan.

Einstein, Albert. 1939. "Albert Einstein to Franklin D. Roosevelt, August 2, 1939," *Atomic Heritage Foundation,* accessed 6/21/2022. https://ahf.nuclearmuseum.org/ahf/key-documents/einstein-szilard-letter/

Future of Life Institute. 2023. "Pause Giant AI Experiments: An Open Letter," *Future of Life Institute,* March 22, 2023. https://futureoflife.org/open-letter/pause-giant-ai-experiments/

Goldman, Sharon. 2023. "Open letter calling for AI 'pause' shines light on fierce debate about risks vs. hype," *VentureBeat,* March 29, 2023. https://venturebeat.com/ai/open-letter-calling-for-ai-pause-shines-light-on-fierce-debate-around-risks-vs-hype/

Haber, Fritz. 1920. "The chemistry in the war," *Speech,* November 11, 1920. University of Stuttgart. https://www.hi.uni-stuttgart.de/wgt/ww-one/Start/Bleed_White/Technology_and_Science/ww1_ger_08_02_01.html

——. 1923. "On the history of gas warfare," *Speech,* October 1, 1923. University of Stuttgart. https://www.hi.uni-stuttgart.de/wgt/ww-one/Start/Bleed_White/Technology_and_Science/ww1_ger_08_02_01.html

Hager, Thomas. 2008. *The Alchemy of Air: A Jewish Genius, A Doomed Tycoon, and the Scientific Discovery that Fed the World but Fueled the Rise of Hitler.* New York: Broadway Books.

Ham, Paul. 2015. "As Hiroshima Smouldered, Our Atom Bomb Scientists Suffered Remorse," *Newsweek,* August 5, 2015. https://www.newsweek.com/hiroshima-smouldered-our-atom-bomb-scientists-suffered-remorse-360125

Herrlich, Peter. 2013. "The Responsibility of the Scientist," *EMBO Reports* 14(9): 759–64.

Hunner, Jon. 2012. *J. Robert Oppenheimer, the Cold War, and the Atomic West.* Norman: University of Oklahoma Press.

Kourany, Janet. 2010. *Philosophy of Science after Feminism.* Oxford University Press.

Kuznick, Peter J. 1989. *Beyond the Laboratory: Scientists as Political Activists in 1930s America.* Chicago: University of Chicago Press.

Lanouette, William, and Bela Silard. 1992. *Genius in the Shadows: A Biography of Leo Szilard the Man Behind the Bomb.* New York: Skyhorse Publishing.

Lübbe, Hermann. 1986. "Scientific Practice and Responsibility." Translated by John N. Kraay. *Facts and Values: Philosophical Reflections from Western and Non-Western Perspectives,* edited by M.C. Doeser and J.N. Kraay. Martinus Nijhoff Publishers.

Markowitz, Gerald, and David Rosner. 2013. *Deceit and Denial: The Deadly Politics of Industrial Pollution.* University of California Press.

McFarland, Micheal J., Matt E. Hauer, and Aaron Reuben. 2022. "Half of US Population exposed to adverse lead levels in early childhood," *Proceedings of the National Academy of Sciences.* https://doi.org/10.1073/pnas.2118631119

McNeill, J.R. 2001. *Something New Under the Sun: An Environmental History of the Twentieth-Century World.* New York: Norton.

Metz, Cade. 2023. "'The Godfather of A.I.' Leaves Google and Warns of Danger Ahead," *The New York Times,* May 4, 2023. https://www.nytimes.com/2023/05/01/technology/ai-google-chatbot-engineer-quits-hinton.html

Moreau, Yves. 2019. "Crackdown on genomic surveillance," *Nature,* December 12, 2019. https://www.nature.com/articles/d41586-019-03687-x

Moulton, F.R. 1938. "Report on Indianapolis Meeting of the AAAS and Associated Societies," *Science* 87: 95–96.

Newsweek. 1947. "Einstein, the Man Who Started It All," *Newsweek,* March 10, 1947.

Noorden, Richard Van. 2020. "The ethical questions that haunt facial-recognition research," *Nature,* November 18, 2020. https://www.nature.com/articles/d41586-020-03187-3

Passano, L. Magruder. 1935. "Ploughing Under the Science Crop," *Science* 81(2089): 46.

Proctor, Robert. 1991. *Value-free Science: Purity and Power in Modern Knowledge.* Cambridge: Harvard University Press.

Raicu, Irina. 2020. "'The Social Dilemma' Is a Half-Missed Opportunity," *Markkula Center for Applied Ethics,* October 26, 2020. https://www.scu.edu/ethics/internet-ethics-blog/the-social-dilemma-is-a-half-missed-opportunity/

Ritter, Steven K. 2008. "The Haber-Bosch Reaction: An Early Chemical Impact on Sustainability," *Chemical & Engineering News,* August 18, 2008. https://cen.acs.org/articles/86/i33/Haber-Bosch-Reaction-Early-Chemical.html

Rudner, Richard. 1953. "The Scientist Qua Scientist Makes Value Judgments," *Philosophy of Science* 20(1): 2.

Smil, Vaclav. 1999. "Detonator of the population explosion," *Nature* (400): 415.

Urey, Harold C. 1948. "Comments on Dr. Bridgman's Article," *Bulletin of the Atomic Scientists* 4(3): 72–73.

Vincent, James. 2017. "Former Facebook exec says social media is ripping apart society," *The Verge,* December 11, 2017. https://www.theverge.com/2017/12/11/16761016/former-facebook-exec-ripping-apart-society

Walker, Mark. 1995. *Nazi Science: Myth, Truth, and the German Atomic Bomb.* New York: Plenum Press.

Wang, Cunrui et al. 2019. "Expression of Concern: Facial Feature Discovery for Ethnicity Recognition," Wiley Interdisciplinary Reviews. *Data Mining and Knowledge Discovery* 9(1): e1278–. https://doi.org/10.1002/widm.1278

Yudkowsky, Eliezer. 2023. "Pausing AI Developments Isn't Enough. We Need to Shut it All Down," *Time*, March 29, 2023. https://time.com/6266923/ai-eliezer-yudkowsky-open-letter-not-enough/

Chapter 3

American Psychiatric Association. 2020. "What is Depression?" *American Psychiatric Association*, October 2020. https://www.psychiatry.org/patients-families/depression/what-is-depression

Angwin, Julia et al. 2016. "Machine Bias," *Propublica*, May 23, 2016. https://www.propublica.org/article/machine-bias-risk-assessments-in-criminal-sentencing

Benner, Katie, Glenn Thrush, and Mike Issac. 2019. "Facebook Engages in Housing Discrimination With Its Ad Practices, U.S. Says," *The New York Times*, March 28, 2019. https://www.nytimes.com/2019/03/28/us/politics/facebook-housing-discrimination.html

Biddle, Justin B. 2022. "On Predicting Recidivism: Epistemic Risk, Tradeoffs, and Values in Machine Learning," *Canadian Journal of Philosophy*: 1–21, doi: 10.1017/can.2020.27

Bolukbasi, Tolga, Kai-Wei Chang, James Zou, Venkatesh Saligrama, and Adam Kalai. 2016. "Man Is to Computer Programmer as Woman Is to Homemaker? Debiasing Word Embeddings." https://doi.org/10.48550/arxiv.1607.06520

Borsboom, Denny, AOJ Cramer, and Annemarie Kalis. 2019. "Brain Disorders? Not really: Why network structures block reductionism in psychopathology research," *Behavioural and Brain Sciences* 42(2):1–63.

Brownlee, John. 2015. "Why Google's Deep Dream A.I. Hallucinates in Dog Faces," *Fast Company*, July 23, 2015. https://www.fastcompany.com/3048941/why-googles-deep-dream-ai-hallucinates-in-dog-faces

Buolamwini, Joy, and Timnit Gebru. 2018. "Gender Shades: Intersectional Accuracy Disparities in Commercial Gender Classification," *Proceedings of Machine Learning Research* 81: 1–15.

Butler, Sarah. 2021. "Uber facing new UK driver claims of racial discrimination," *The Guardian*, October 6, 2021. https://www.theguardian.com/technology/2021/oct/06/uber-facing-new-uk-driver-claims-of-racial-discrimination

Castelvecchi, Davide. 2020. "Is facial recognition too biased to be let loose?" *Nature*, November 18, 2020. https://www.nature.com/articles/d41586-020-03186-4

Chouldechova, Alexandra. 2017. "Fair Prediction with Disparate Impact: A Study of Bias in Recidivism Prediction Instruments," *Big Data* 5(2). https://doi.org/10.1089/big.2016.0047

Crenshaw, Kimberle. 1989. "Demarginalizing the Intersection of Race and Sex: A Black Feminist Critique of Antidiscrimination Doctrine, Feminist Theory and Antiracist Politics," *University of Chicago Legal Forum* 1: 139–68.

Crockford, Kade. 2020. "How is Face Recognition Surveillance Technology Racist?" *ACLU*, June 16, 2020. https://www.aclu.org/news/privacy-technology/how-is-face-recognition-surveillance-technology-racist

Dastin, Jeffrey. 2018. "Amazon Scraps Secret AI Recruiting Tool that

Showed Bias Against Women,"
Reuters, October 10, 2018. https://
www.reuters.com/article/us-amazon-
com-jobs-automation-insight-
idUSKCN1MK08G

Dewey, John. 1998. "Context and
Thought," *The Essential Dewey Vol.
1*, edited by L.A. Hickman and T.M.
Alexander. Bloomington: Indiana
University Press.

Duhem, Pierre. 1954. *The Aim and
Structure of Physical Theory*. Princeton:
Princeton University Press.

Ewart, Asia. 2020. "The Shirley Card:
Racial Photographic Bias through
Skin Tone," *Shutterstock*, July 30, 2020.
https://www.shutterstock.com/blog/
shirley-card-racial-photographic-bias

Gebru, Timnit. 2020. "Race and Gender,"
The Oxford Handbook of Ethics of AI,
edited by Markus D. Dubber, Frank
Pasquale, and Sunit Das. Oxford:
Oxford University Press.

Gebru, Timnit, Jamie Morgenstern,
Briana Vecchione, Jennifer Wortman
Vaughan, Hanna Wallach, Hal
Daumé III, and Kate Crawford.
2021. "Datasheets for Datasets,"
Communications of the ACM 64(12):
86–92.

Hao, Karen. 2019. "AI is sending people
to jail – and getting it wrong," *MIT
Technology Review*, January 21, 2019.
https://www.technologyreview.
com/2019/01/21/137783/
algorithms-criminal-justice-ai/

Harwell, Drew. 2019. "Federal study
confirms racial bias of many facial-
recognition systems, casts doubt
on their expanding use," *The
Washington Post*, December 19,
2019. https://www.washingtonpost.
com/technology/2019/12/19/
federal-study-confirms-racial-bias-
many-facial-recognition-systems-
casts-doubt-their-expanding-use/

Horwitz, Jeff. 2021. "Facebook Algorithm
Shows Gender Bias in Job Ads,
Study Finds," *The Wall Street Journal*,
April 9, 2021. https://www.wsj.com/
articles/facebook-shows-men-and-
women-different-job-ads-study-
finds-11617969600

IBM. 2022. "Watson OpenScale On
Cloud Pak for Data," *IBM*, last
updated October 26, 2022. https://
www.ibm.com/docs/en/cloud-paks/
cp-data/3.5.0?topic=services-watson-
openscale

IBM Developer Staff. 2018. "AI Fairness
360," *IBM*, November 14, 2018. https://
www.ibm.com/opensource/open/
projects/ai-fairness-360/

Intemann, Kristen. 2001. "Science and
Values: Are Value Judgments Always
Irrelevant to the Justification of
Scientific Claims?" *Philosophy of
Science* 68: S506–518.

Ito, Joi. 2019. "Supposedly 'fair' algorithms
can perpetuate discrimination,"
MIT Media Lab, February 5, 2019.
https://www.media.mit.edu/articles/
supposedly-fair-algorithms-can-
perpetuate-discrimination/

Kansal, Tushar. 2005. "Racial Disparity
in Sentencing," *Open Society
Foundations*, January 2005. https://
www.opensocietyfoundations.
org/publications/
racial-disparity-sentencing

Kayser-Bril, Nicolas. 2020. "Female
historians and male nurses do
not exist, Google Translate tells
its European users," *Algorithm
Watch*, September 17, 2020.
https://algorithmwatch.org/en/
google-translate-gender-bias/

Kearns, Michael, and Aaron Roth. 2020.
*The Ethical Algorithm: The Science
of Socially Aware Algorithm Design*.
Oxford University Press.

Kleinberg, Jon, Sendhil Mullainathan, and
Manish Raghavan. 2016. "Inherent
Trade-Offs in the Fair Determination
of Risk Scores." https://doi.
org/10.48550/arxiv.1609.05807

Larson, Jeff, Surya Mattu, Lauren
Kirchner, and Julia Angwin. 2016.

"How We Analyzed the COMPAS Recidivism Algorithm," *Propublica*, May 23, 2016. https://www.propublica.org/article/how-we-analyzed-the-compas-recidivism-algorithm

Ledford, Heidi. 2019. "Millions of black people affected by racial bias in health-care algorithms," *Nature*, October 24, 2019. https://www.nature.com/articles/d41586-019-03228-6

Lewis, Sarah. 2019. "The Racial Bias Built into Photography," *The New York Times*, April 25, 2019. https://www.nytimes.com/2019/04/25/lens/sarah-lewis-racial-bias-photography.html

Longino, Helen. 1990. *Science as Social Knowledge*. Princeton: Princeton University Press.

——. 2001. *The Fate of Knowledge*. Princeton: Princeton University Press.

Mitchell, Tom M. 1997. *Machine Learning*. McGraw-Hill Science.

Munn, Nathan. 2020. "Police Across Canada Are Using Predictive Policing Algorithms, Report Finds," *Vice News*, September 1, 2020. https://www.vice.com/en/article/k7q55x/police-across-canada-are-using-predictive-policing-algorithms-report-finds

Nagel, Thomas. 1989. *The View from Nowhere*. New York: Oxford University Press.

Nandi, Anirban, and Aditya Kumar Pal. 2022. *Interpreting Machine Learning Models*. New York: Apress.

Nemeroff, Charles. 1998. "The Neurobiology of Depression," *Scientific American*, June 1, 1998: 42–49.

Obermeyer, Ziad, Brian Powers, Christine Vogeli, and Sendhil Mullainathan. 2019. "Dissecting Racial Bias in an Algorithm Used to Manage the Health of Populations," *Science (American Association for the Advancement of Science)* 366(6464): 447–53. https://doi.org/10.1126/science.aax2342

O'Neil, Cathy. 2016. *Weapons of Math Destruction: How Big Data Increases Inequality and Threatens Democracy*. New York: Crown Publishing.

Revhavi, M. Marit, and Sonja B. Starr. 2014. "Racial Disparity in Federal Criminal Sentences," *Journal of Political Economy* 122(6): 1320–54.

Roth, Lorna. 2009. "Looking at Shirley, the Ultimate Norm: Color Balance, Image Technologies, and Cognitive Equality," *Canadian Journal of Communication* 34(1): 111–36.

Sankin, Aaron, Dhruv Mehrotra, Surya Mattu, Dell Cameron, Annie Gilbertson, Daniel Lempres, and Josh Lash. 2021. "Crime Prediction Software Promised to be Free of Biases. New Data Shows It Perpetuates Them," *Gizmodo*, December 2, 2021. https://gizmodo.com/crime-prediction-software-promised-to-be-free-of-biases-1848138977

Shueh, Jason. 2016. "White House Challenges Artificial Intelligence Experts to Reduce Incarceration Rates," *Government Technology*, June 07, 2016. https://www.govtech.com/computing/white-house-challenges-artificial-intelligence-experts-to-reduce-incarceration-rates.html

Singer, Natasha. 2012. "E-Tailer Customization: Convenient or Creepy?" *The New York Times*, June 20, 2012. https://www.nytimes.com/2012/06/24/technology/e-tailer-customization-whats-convenient-and-whats-just-plain-creepy.html

Singh, Jatinder, Ian Walden, Jon Crowcroft, and Jean Bacon. 2016. "Responsibility and Machine Learning: Part of a Process," *SSRN*, October 27, 2016. http://dx.doi.org/10.2139/ssrn.2860048

Smith, Craig S. 2019. "Dealing With Bias in Artificial Intelligence," *The New York Times*, November 19, 2019. https://www.nytimes.com/2019/11/19/technology/artificial-intelligence-bias.html

Spector-Bagdady, Kayte, Jenna Weins, and Melissa Creary. 2021. "Study shows bias can creep into medical databanks that drive precision health and clinical AI," *University of Michigan Institute for Healthcare Policy & Innovation*, December 7, 2021. https://ihpi.umich.edu/news/study-shows-how-bias-can-creep-medical-databanks-drive-precision-health-and-clinical-ai

Vartan, Starre. 2019. "Racial Bias Found in a Major Health Care Risk Algorithm," *Scientific American*, October 24, 2019. https://www.scientificamerican.com/article/racial-bias-found-in-a-major-health-care-risk-algorithm/

Verma, Sahil, and Julia Rubin. 2018. "Fairness Definitions Explained," in 2018 IEEE/ACM International Workshop on Software Fairness (FairWare), 1–7. ACM. https://doi.org/10.1145/3194770.3194776

Vincent, James. 2021. "Automatic gender recognition tech is dangerous, say campaigners: It's time to ban it," *The Verge*, April 14, 2021. https://www.theverge.com/2021/4/14/22381370/automatic-gender-recognition-sexual-orientation-facial-ai-analysis-ban-campaign

Wong, Julia Carrie. 2019. "The viral selfie app ImageNet Roulette seemed fun – until it called me a racist slur," *The Guardian*, September 18, 2019. https://www.theguardian.com/technology/2019/sep/17/imagenet-roulette-asian-racist-slur-selfie

Chapter 4

Aschwanden, Christie. 2015. "Science Isn't Broken," *FiveThirtyEight*, August 19, 2015. https://fivethirtyeight.com/features/science-isnt-broken/#part1

——. 2016. "You Can't Trust What You Read About Nutrition," *FiveThirtyEight*, January 6, 2016. https://fivethirtyeight.com/features/you-cant-trust-what-you-read-about-nutrition/

Balogh, Shannen, and Carter Johnson. 2021. "AI can help reduce inequity in credit access, but banks will have to trade off fairness for accuracy – for now," *Business Insider*, June 30, 2021. https://www.businessinsider.com/ai-lending-risks-opportunities-credit-decisioning-data-inequity-2021-6

Billott, Rob. 2020. "The poison found in everyone, even unborn babies – and who is responsible for it," *The Guardian*, December 17, 2020. https://www.theguardian.com/commentisfree/2020/dec/17/dark-waters-pfas-ticking-chemical-time-bomb-in-your-blood

Bohannon, John. 2016. "About 40% of economics experiments fail replication survey," *Science*, March 3, 2016. https://www.science.org/content/article/about-40-economics-experiments-fail-replication-survey-rev2

Brendel, Wieland, and Matthias Bethge. 2019. "Approximating CNNs with Bag-of-Local-Features Models Works Surprisingly Well on ImageNet." https://doi.org/10.48550/arxiv.1904.00760

Burt, Andrew. 2019. "The AI Transparency Paradox," *Harvard Business Review*, December 13, 2019. https://hbr.org/2019/12/the-ai-transparency-paradox

Clifford, W.K. 1879. "Right and Wrong: The Scientific Ground of Their Distinction," *Lectures and Essays Volume 2*, edited by Leslie Stephen and Frederick Pollock, 124–76. London: MacMillan and Co.

——. 1879. "The Ethics of Belief," *Lectures and Essays Volume 2*, edited by Leslie Stephen and Frederick Pollock, 177–211. London: MacMillan and Co.

Dattner, Ben, Tomas Chomorro-Premuzic, Richard Buchband, and Lucinda Schettler. 2019. "The Legal and Ethical Implications of Using

AI in Hiring," *Harvard Business Review*, April 25, 2019. https://hbr.org/2019/04/the-legal-and-ethical-implications-of-using-ai-in-hiring

Dignum, Virginia. 2020. *Responsible Artificial Intelligence: How to Develop and Use AI in a Responsible Way*. Umeå: Springer International Publishing.

Ellin, Abby. 2012. "Woman Sues Over Personality Test Job Rejection," ABC News, October 1, 2012. https://abcnews.go.com/Business/personality-tests-workplace-bogus/story?id=17349051

Evalueserve. 2022. "How AI is Transforming the Auto Insurance Industry with Charles Boyle," *Evalueserve*, accessed July 25, 2023. https://www.evalueserve.com/podcast/how-ai-and-data-analytics-are-transforming-the-auto-insurance-industry-with-charles-boyle/

Felder, Ryan Marshall. 2021. "Coming to Terms with the Black Box Problem: How to Justify AI Systems in Health Care," *Hastings Center Report* 4: 38–45.

Gill, Lisa L. 2021. "More Than a Third of Volunteers in a Consumer Reports Study Found Errors in Their Credit Reports," *Consumer Reports*, updated June 11, 2021. https://www.consumerreports.org/credit-scores-reports/consumers-found-errors-in-their-credit-reports-a6996937910/

Head, Megan L., Luke Holman, Rob Lanfear, Andrew T. Kahn, and Michael D. Jennions. 2015. "The Extent and Consequences of p-Hacking in Science," *PLoS Biology* 13(3): e1002106–e1002106. https://doi.org/10.1371/journal.pbio.1002106

Ioannidis, John P.A. 2005. "Why Most Published Research Findings are False," *PLoS Med* 2(8): 696–701.

Le Quoc, V. 2013. "Building High-Level Features Using Large Scale Unsupervised Learning." In *2013 IEEE International Conference on Acoustics, Speech and Signal Processing*, 8595–98.

IEEE. https://doi.org/10.1109/ICASSP.2013.6639343

Lytton, Charlotte. 2024. "AI hiring tools may be filtering out the best job applicants," *BBC*, February 16, 2024. https://www.bbc.com/worklife/article/20240214-ai-recruiting-hiring-software-bias-discrimination

Marche, Stephen. 2022. "The College Essay is Dead," *The Atlantic*, December 6, 2022. https://www.theatlantic.com/technology/archive/2022/12/chatgpt-ai-writing-college-student-essays/672371/

Martin, G.N., and Richard M. Clarke. 2017. "Are Psychology Journals Anti-Replication? A Snapshot of Editorial Practices," *Frontiers in Psychology* 8: 523. https://doi.org/10.3389/fpsyg.2017.00523

Meinert, Dori. 2015. "What Do Personality Tests Really Reveal?" *SHRM*, June 1, 2015. https://www.shrm.org/hr-today/news/hr-magazine/pages/0615-personality-tests.aspx

Milli, Smitha, Ludwig Schmidt, Anca Dragan, and Moritz Hardt. 2019. "Model Reconstruction from Model Explanations," in *Proceedings of the Conference on Fairness, Accountability, and Transparency*, 1–9. ACM. https://doi.org/10.1145/3287560.3287562

Nilesh, Christopher. 2023. "An Indian politician says scandalous audio clips are AI deepfakes. We had them tested." *Rest of World*, July 5, 2023. https://restofworld.org/2023/indian-politician-leaked-audio-ai-deepfake/

Nix, Naomi, and Elizabeth Dwoskin. 2022. "Justice Department and Meta settle landmark housing discrimination case," *The Washington Post*, June 21, 2022. https://www.washingtonpost.com/technology/2022/06/21/facebook-doj-discriminatory-housing-ads/

O'Neil, Cathy. 2018. "Personality Tests Are Failing American Workers," *Bloomberg*, January 18, 2018. https://

www.bloomberg.com/view/ articles/2018-01-18/personality-tests-are-failing-american-workers

Open Science Collaboration. 2015. "Estimating the reproducibility of psychological science," *Science* 349(6251). https://doi.org/10.1126/ science.aac4716

Pappas, George. 2014. "Internalist vs. Externalist Conceptions of Justification," *Stanford Encyclopedia of Philosophy*. https://plato.stanford.edu/ entries/justep-intext/

Prospect AI. 2019. "The Most Effective Pre-employment Assessment," *Medium*, June 10, 2019. https:// medium.com/perspectai/the-most-effective-pre-employment-assessment-6ff985fc6bf2

Renzulli, Kerri Anne. 2019. "75% of resumes never read by a human – here's how to make sure yours does," *Yahoo! Finance*, February 28, 2019. https://finance.yahoo. com/news/75-resumes-never-read-human-174855340.html

Resnick, Brian. 2017. "What a nerdy debate about p-values shows about science – and how to fix it," *Vox*, July 31, 2017. https:// www.vox.com/science-and-health/2017/7/31/16021654/p-values-statistical-significance-redefine-0005

——. 2019. "800 Scientists say it's time to abandon 'statistical significance'," *Vox*, March 22, 2019. https://www.vox. com/latest-news/2019/3/22/18275913/ statistical-significance-p-values-explained

Ribeiro, Marco Tulio, Sameer Singh, and Carlos Guestrin. 2016. "'Why Should I Trust You?': Explaining the Predictions of Any Classifier," *Proceedings of the 22nd ACM SIGKDD International Conference on Knowledge Discovery and Data Mining*, 1135–44. ACM. https://doi. org/10.1145/2939672.2939778

Rich, Nathaniel. 2016. "The Lawyer Who Became Dupont's Worst Nightmare," *The New York Times*, January 6, 2016. https://www.nytimes.com/2016/01/10/ magazine/the-lawyer-who-became-duponts-worst-nightmare.html

Rudin, Cynthia. 2019. "Stop explaining Black Box Machine Learning Models for High Stakes Decisions and Use Interpretable Models Instead," *Natural Machine Intelligence* 1(5): 206–15.

Shane, Janelle. 2020. "When data is messy," *AI Weirdness*, July 3, 2020. https://www.aiweirdness.com/ when-data-is-messy-20-07-03/

Shokri, Reza, Martin Strobel, and Yair Zick. 2019. "On the Privacy Risks of Model Explanations." https://doi. org/10.48550/arxiv.1907.00164

Singer, Natasha. 2012. "Secret E-Score Chart Consumers' Buying Power," *The New York Times*, August 18, 2012. https://www.nytimes. com/2012/08/19/business/electronic-scores-rank-consumers-by-potential-value.html

Singh, Ishaan. n.d. "AI-Based Algorithms for Credit Scoring in Personal Loans," *Medium*, accessed 7/25/2023. https://medium.com/@ ishaansinghh96/ai-based-algorithms-for-credit-scoring-in-personal-loans-6168128b5b07

Slack, Dylan, Sophie Hilgard, Emily Jia, Sameer Singh, and Himabindu Lakkaraju. 2020. "Fooling LIME and SHAP: Adversarial Attacks on Post Hoc Explanation Methods," in *Proceedings of the AAAI/ACM Conference on AI, Ethics, and Society*, 180–86. ACM. https://doi. org/10.1145/3375627.3375830

Staley, Kent W. 2017. "Decisions, Decisions: Inductive Risk and the Higgs Boson," in *Exploring Inductive Risk: Case Studies of Values in Science*, edited by Kevin C. Elliott and Ted Richards. New York: Oxford University Press.

Tomsett, Richard, Dave Braines, Dan Harborne, Alun Preece, and Supriyo Chakraborty. 2018. "Interpretable to Whom? A Role-Based Model for Analyzing Interpretable Machine Learning Systems." https://doi.org/10.48550/arxiv.1806.07552

Tracy, Ryan, and Georgia Wells. 2022. "TikTok's Secret Sauce Poses Challenge for U.S. Oversight, Researchers Say," *The Wall Street Journal*, February 8, 2023. https://www.wsj.com/articles/tiktoks-secret-sauce-poses-challenge-for-u-s-oversight-researchers-say-11675818735

Turque, Bill. 2012. "'Creative ... motivating' and fired," *The Washington Post*, March 6, 2012. https://www.washingtonpost.com/local/education/creative--motivating-and-fired/2012/02/04/gIQAwzZpvR_story.html

Uribe, Francisco Mejia. 2018. "Believing without evidence is always morally wrong," edited by Nigel Warburton. *AEON*, November 5, 2018. https://aeon.co/ideas/believing-without-evidence-is-always-morally-wrong

VentureBeat. 2021. "When AI Flags the Ruler not the Tumour," *VentureBeat*, March 25, 2021. https://venturebeat.com/business/when-ai-flags-the-ruler-not-the-tumor-and-other-arguments-for-abolishing-the-black-box-vb-live/

Wolkovich, E.M. 2024. "'Obviously ChatGPT' – how reviewers accused me of scientific fraud," *Nature*, February 05, 2024. https://www.nature.com/articles/d41586-024-00349-5

Zednik, Carlos. 2021. "Solving the Black Box Problem: A Normative Framework for Explainable Artificial Intelligence," *Philosophy & Technology* 2021(34): 265–88.

Chapter 5

Addams, Jane. 1902. *Democracy and Social Ethics*. New York: The MacMillan Company.

Akinwotu, Emmanuel. 2021. "Facebook's role in Myanmar and Ethiopia under new scrutiny," *The Guardian*, October 7, 2021. https://www.theguardian.com/technology/2021/oct/07/facebooks-role-in-myanmar-and-ethiopia-under-new-scrutiny

Al Jazeera. 2018. "Myanmar: Security forces face 'action' over killings," *Al Jazeera*, February 11, 2018. https://www.aljazeera.com/news/2018/2/11/myanmar-security-forces-face-action-over-killings

Arandjelovic, Ognjen. 2021. "AI, Democracy, and the Importance of Asking the Right Questions," *The AI Ethics Journal* 2(2). https://doi.org/10.47289/aiej20210910

Barocas S. 2012. "The price of precision: voter microtargeting and its potential harms to the democratic process," in *Proceedings of the First Edition Workshop on Politics, Elections and Data*, 31–36. ACM. https://doi.org/10.1145/2389661.2389671

Bontridder, Noémi, and Yves Poullet. 2021. "The Role of Artificial Intelligence in Disinformation," *Data & Policy*, 3: e32. doi:10.1017/dap.2021.20

Borgesius, F., J. Zuiderveen, J. Möller, S. Kruikemeier, R. Ó Fathaigh, K. Irion, T. Dobber, B. Bodo, and C. de Vreese. 2018. "Online Political Microtargeting: Promises and Threats for Democracy," *Utrecht Law Review* 14(1): 82–96.

Brennan, Jason. 2011. "The Right to a Competent Electorate," *The Philosophical Quarterly* 61(245): 699–724.

Christiano, Thomas. 2022. "Algorithms, Manipulation, and Democracy," *Canadian Journal of Philosophy* 52(1): 109–24. https://doi.org/10.1017/can.2021.29

Cofessore, Nicholas. 2018. "Cambridge Analytica and Facebook: The Scandal and the Fallout So Far," *The New York Times*, April 4, 2018. https://www.nytimes.com/2018/04/04/us/politics/cambridge-analytica-scandal-fallout.html

Cost, Ben, Asia Grace, Marisa Dellatto, Eric Hegedus, and Sophie Gardiner. 2023. "The 25 craziest TikTok challenges so far – and the ordeals they've caused," *New York Post*, updated July 27, 2023. https://nypost.com/article/craziest-tiktok-challenges-so-far/

Dea, Shannon, and Matthew S.W. Silk. "Sympathetic Knowledge and the Scientific Attitude: Classic Pragmatist Resources for Feminist Social Epistemology," in *The Routledge Handbook of Social Epistemology*, edited by Miranda Fricker, Peter J. Graham, David Henderson, and Nikolai J.L.L. Pederson. New York: Routledge.

Dearden, Lizzie. 2019. "Police stop people for covering their faces from facial recognition camera then fine man £90 after he protested," *Independent*, January 31, 2019. https://www.independent.co.uk/news/uk/crime/facial-recognition-cameras-technology-london-trial-met-police-face-cover-man-fined-a8756936.html

Dewey, John. 1916. *Democracy and Education*. New York: The MacMillan Company.

——. 1989. *Freedom and Culture*. Amherst: Prometheus Books.

——. 2016. *The Public and Its Problems*, edited by Melvin L. Rogers. Athens: Swallow Press.

Elshtain, Jean Bethke. 2002. *Jane Addams and the Dream of American Democracy*. New York: Basic Books.

Gorton, William A. "Manipulating Citizens: How Political Campaigns' Use of Behavioral Science Harms Democracy," *New Political Science* 38(1): 61–80.

Hajli, Nick, Usman Saeed, Mina Tajvidi, and Farid Shirazi. 2022. "Social Bots and the Spread of Disinformation in Social Media: The Challenges of Artificial Intelligence," *British Journal of Management* 33: 1238–53.

Heath, Ryan. 2023. "Democracy isn't ready for its AI test," *AXIOS*, May 11, 2023. https://www.axios.com/2023/05/11/democracy-ai-artificial-intelligence-2024-elections

Hooker, Lucy. 2017. "Have you been nudged?" *BBC News*, October 9, 2017. https://www.bbc.com/news/business-41549533

Kang, Cecilia, and Adam Satariano. 2023. "As A.I. Booms, Lawmakers Struggle to Understand the Technology," *The New York Times*, March 3, 2023. https://www.nytimes.com/2023/03/03/technology/artificial-intelligence-regulation-congress.html

Kertysova, Katarina. 2018. "Artificial Intelligence and Disinformation," *Security and Human Rights* 29: 55–81.

Knight, Louise W. 2005. *Citizen: Jane Addams and the Struggle for Democracy*. Chicago: University of Chicago Press.

Lippmann, Walter. 1919. "The Basic Problem of Democracy," *The Atlantic*, November 1919. https://www.theatlantic.com/magazine/archive/1919/11/the-basic-problem-of-democracy/569095/

——. 1922. *Public Opinion*. New York: The MacMillian Company.

——. 1993. *The Phantom Public*. New Brunswick: Transaction Publishers.

Lyons, Kim. 2020. "An Indian politician used AI to translate his speech into other languages to reach more voters," *The Verge*, February 18, 2020. https://www.theverge.com/2020/2/18/21142782/india-politician-deepfakes-ai-elections

Menand, Louis. 2001. *The Metaphysical Club: A Story of Ideas in America*. New York: Farrar, Straus and Giroux.

Menn, Joseph. 2023. "Russians boasted that just 1% of fake social profiles are caught, leak shows," *The Washington Post*, April 16, 2023. https://www.washingtonpost.com/technology/2023/04/16/russia-disinformation-discord-leaked-documents/

Mozur, Paul. 2018. "A Genocide Incited on Facebook, With Post from Myanmar's Military," *The New York Times*, October 15, 2018. https://www.nytimes.com/2018/10/15/technology/myanmar-facebook-genocide.html

Popli, Nik. 2021. "The 5 Most Important Revelations From the 'Facebook Papers,'" *TIME*, October 25, 2021. https://time.com/6110234/facebook-papers-testimony-explained/

Romero, Alberto. 2022. "China Has Pioneered a Law To Empower People Over Algorithms," *Medium*, March 8, 2022. https://onezero.medium.com/china-has-pioneered-a-law-to-empower-people-over-algorithms-70b29ba6285f

Runciman, David. 2018. "Why replacing politicians with experts is a reckless idea," *The Guardian*, May 1, 2018. https://www.theguardian.com/news/2018/may/01/why-replacing-politicians-with-experts-is-a-reckless-idea

Saletta, Morgan. n.d. "Of Zombies and Trump: Changing the World with Microtargeting," *Hippo Reads*, accessed May 29, 2023. https://hipporeads.com/of-zombies-and-trump-changing-the-world-with-microtargeting/

Shaffer, Kris. 2019. *Data vs. Democracy: How Big Data Algorithms Shape Opinion and Alter the Course of History*. New York: Apress.

Silverberg, David. 2023. "Could AI Swamp Social Media with Fake Accounts?" BBC News, February 14, 2023. https://www.bbc.com/news/business-64464140

Solon, Olivia. 2020. "Amazon issues 12-month ban on police use of facial recognition," *NBC News*, June 10, 2020. https://www.nbcnews.com/tech/security/amazon-issues-12-month-ban-police-use-facial-recognition-n1229601

Thaler, Richard H., and Cass R. Sunstein. 2008. *Nudge: Improving Decisions About Health, Wealth, and Happiness*. New York: Penguin Books.

Vincent, James. 2020. "NYPD used facial recognition to track down Black Lives Matter activist," *The Verge*, August 18, 2020. https://www.theverge.com/2020/8/18/21373316/nypd-facial-recognition-black-lives-matter-activist-derrick-ingram

Walker, Mason, and Katerina Eva Matsa. 2021. "News Consumption Across Social Media in 2021," *Pew Research Center*, September 20, 2021. https://www.pewresearch.org/journalism/2021/09/20/news-consumption-across-social-media-in-2021/

Yeung, Karen. 2017. "'Hypernudge': Big Data as a Mode of Regulation by Design," *Information, Communication & Society* 20(1): 118–36. https://doi.org/10.1080/1369118X.2016.1186713

Chapter 6

Allcott, Hunt, Matthew Gentzkow, and Lena Song. 2022. "Digital Addiction," *American Economic Review* 112(7): 2424–63.

Allen, Felix. 2020. "DYING FOR LIKES Dark Truth of social media as US pre-teen girl suicides soar 150% & self-harm TRIPLES, Netflix's Social Dilemma Reveals," *The U.S. Sun*, September 17, 2020. https://www.the-sun.com/news/1487147/

social-media-suicides-self-harm-netflix-social-dilemma/

Allyn, Bobby. 2022. "The Google engineer who sees company's AI as 'sentient' thinks a chatbot has a soul," *NPR*, June 16, 2022. https://www.npr.org/2022/06/16/1105552435/google-ai-sentient

American Psychiatric Association. 2013. *Diagnostic and statistical manual of mental disorders*, 5th ed. Washington, DC: American Psychiatric Association.

Andreassen, Cecilie Schou. 2015. "Online Social Network Site Addiction: A Comprehensive Review," *Current Addiction Reports* 2(2): 175–84. https://doi.org/10.1007/s40429-015-0056-9

Aristotle. 2013. *Aristotle's "Politics,"* 2nd ed., translated by Carnes Lord. Chicago: University of Chicago Press.

Basen, Ira. 2018. "You can't stop checking your phone because Silicon Valley designed it that way," *CBC Radio*, September 14, 2018. https://www.cbc.ca/radio/sunday/the-sunday-edition-september-16-2018-1.4822353/you-can-t-stop-checking-your-phone-because-silicon-valley-designed-it-that-way-1.4822360

Basiousny, Angie. 2023. "ChatGPT Passed an MBA Exam. What's Next?" *Knowledge at Wharton*, January 31, 2023. https://knowledge.wharton.upenn.edu/podcast/wharton-business-daily-podcast/chatgpt-passed-an-mba-exam-whats-next/

Bedingfield, Will. 2022. "The Bruce Willis Deepfake is Everyone's Problem," *Wired*, October 17, 2022. https://www.wired.co.uk/article/bruce-willis-deepfake-rights-law

Benn, Stanley I. 1984. "Privacy, Freedom, and Respect for Persons," in *Philosophical Dimensions of Privacy: An Anthology*, edited by Ferdinand David Schoeman, 223–244. Cambridge: Cambridge University Press. https://10.1017/CBO9780511625138.009

Bhargava, Vikram R., and Manuel Velasquez. 2021. "Ethics of the Attention Economy: The Problem of Social Media Addiction," *Business Ethics Quarterly* 31(3): 321–59. https://10.1017/beq.2020.32

Blistein, Jon. 2023. "Autonomous AI-Generated Artwork Cannot be Copyrighted, Judge Rules," *Rolling Stone*, August 18, 2023. https://www.rollingstone.com/culture/culture-news/ai-generated-artwork-not-protected-copyright-law-1234809089/

Brändlin, Anne-Sophie. 2015. "The Impact of Photoshop," *DW*, February 27, 2015. https://www.dw.com/en/how-25-years-of-photoshop-changed-our-perception-of-reality/a-18284410

Brewster, Thomas. 2021. "Fraudsters Cloned Company Director's Voice in $35 Million Heist, Police Find," *Forbes*, October 14, 2021. https://www.forbes.com/sites/thomasbrewster/2021/10/14/huge-bank-fraud-uses-deep-fake-voice-tech-to-steal-millions

Brown, Annie. 2021. "Making The YouTube Algorithm Less Elusive With The Help of Gregory Chase, A Creator With 10M+ Subscribers," *Forbes*, July 13, 2021. https://www.forbes.com/sites/anniebrown/2021/07/13/making-the-youtube-algorithm-less-elusive-with-the-help-of-gregory-chase-a-creator-with-10m-subscribers/?sh=7c4916bad681

Cabral, Jaclyn. 2011. "Is Generation Y Addicted to Social Media?" *Elon Journal of Undergraduate Research in Communications* 2(1): 5–14.

CBC Arts. 2023. "Meet Alsis, the band fronted by an AI Liam Gallager," *CBC*, April 14, 2023. https://www.cbc.ca/arts/commotion/meet-aisis-the-band-fronted-by-an-ai-liam-gallagher-1.6820734

Cho, Winston. 2023. "Actor's AI Protections Are a Step Forward, But There's Reason to Worry," *The*

Hollywood Reporter, November 14, 2023. https://www.hollywoodreporter.com/business/business-news/where-sag-aftra-deal-may-miss-ai-protections-1235646988/

Clark, Mitchell. 2022. "The engineer who claimed a Google AI is sentient has been fired," *The Verge*, July 22, 2022. https://www.theverge.com/2022/7/22/23274958/google-ai-engineer-blake-lemoine-chatbot-lamda-2-sentience

Dalton, Andrew, and the Associated Press. 2023. "Writers strike: Why A.I. is such a hot-button issue in Hollywood's labor battle with SAG-AFTRA," *Fortune*, July 24, 2023. https://fortune.com/2023/07/24/sag-aftra-writers-strike-explained-artificial-intelligence/

Das, Anaisha, Neetra Chakraborty, and Srishreya Arunsaravanakumar. 2022. "AI-generated art raises ethical concerns," *The Wildcat Tribune*, October 6, 2022. https://thewildcattribune.com/15530/ae/ai-generated-art-raises-ethical-concerns/

Davis, Lizzie. 2017. "Emma Watson's voice is too auto-tuned in Beauty and the Beast, says professional soprano," *Classic FM*, last updated April 18, 2018. https://www.classicfm.com/discover-music/periods-genres/film-tv/emma-watson-singing-beauty-and-the-beast/

De Ruiter, Adrienne. 2021. "The Distinct Wrong of Deepfakes," *Philosophy & Technology* 34: 1311–32.

Edwards, Benj. 2023. "Why ChatGPT and Bing Chat are so good at making things up," *ArsTechnica*, April 6, 2023. https://arstechnica.com/information-technology/2023/04/why-ai-chatbots-are-the-ultimate-bs-machines-and-how-people-hope-to-fix-them/

Flint, Joe. 2015. "In L.A., One Way to Beat Traffic Runs Into Backlash," *The Wall Street Journal*, November 13, 2015. https://www.wsj.com/articles/in-l-a-one-way-to-beat-traffic-runs-into-backlash-1447469058

Fricker, Miranda. 2007. *Epistemic Injustice: Power and the Ethics of Knowing*. Oxford: Oxford University Press.

Gault, Matthew. 2022. "An AI-Generated Artwork Won First Place at a State Fair Fine Arts Competition, and Artists Are Pissed," *Vice*, August 31, 2022. https://www.vice.com/en/article/bvmvqm/an-ai-generated-artwork-won-first-place-at-a-state-fair-fine-arts-competition-and-artists-are-pissed

Geher, Glenn. 2023. "ChatGPT, Artificial Intelligence, and the Future of Writing," *Psychology Today*, February 6, 2023. https://www.psychologytoday.com/ca/blog/darwins-subterranean-world/202301/chatgpt-artificial-intelligence-and-the-future-of-writing

Goodwin, Danny. 2023. "Google warns against content pruning as CNET deletes thousands of pages," *Search Engine Land*, August 9, 2023. https://searchengineland.com/google-warns-against-content-pruning-as-cnet-deletes-thousands-of-pages-430509

Haidt, Jonathan. 2021. "The Dangerous Experiment in Teen Girls," *The Atlantic*, November 21, 2021. https://www.theatlantic.com/ideas/archive/2021/11/facebooks-dangerous-experiment-teen-girls/620767/

Haidt, Jonathan, and Jean M. Twenge. 2021. "This is Our Chance to Pull Teenagers Out of the Smartphone Trap," *The New York Times*, July 31, 2021. https://www.nytimes.com/2021/07/31/opinion/smartphone-iphone-social-media-isolation.html

Hancock, Jay. 2015. "Workplace wellness programs put employee privacy at risk," *CNN Health*, September 28, 2015. https://www.cnn.com/2015/09/28/health/workplace-wellness-privacy-risk-exclusive/index.html

Hao, Karen. 2019. "The biggest threat to deepfakes isn't the deepfakes themselves," *MIT Technology*

Review, October 19, 2019. https://www.technologyreview.com/2019/10/10/132667/the-biggest-threat-of-deepfakes-isnt-the-deepfakes-themselves/

———. 2021. "Deepfake porn is ruining women's lives. Now the law may finally ban it," *MIT Technology Review*, February 12, 2021. https://www.technologyreview.com/2021/02/12/1018222/deepfake-revenge-porn-coming-ban/

Heilweil, Rebecca. 2022. "Why Silicon Valley is fertile ground for obscure religious beliefs," *Vox*, June 30, 2022. https://www.vox.com/recode/2022/6/30/23188222/silicon-valley-blake-lemoine-chatbot-eliza-religion-robot

Hetrick, Christian. 2022. "Art Created by Artificial Intelligence Can't Be Copyrighted, US Agency Rules," *Dot. LA*, February 21, 2022. https://dot.la/creative-machines-ai-art-2656764050.html

Hill, Kashmir. 2012. "How Target Figured Out a Teen Girl Was Pregnant Before Her Father Did," *Forbes*, February 16, 2012. https://www.forbes.com/sites/kashmirhill/2012/02/16/how-target-figured-out-a-teen-girl-was-pregnant-before-her-father-did/?sh=3b7976446668

Hunt, Melissa G., Rachel Marx, Courtney Lipson, and Jordyn Young. 2018. "No More FOMO: Limiting Social Media Decreases Loneliness and Depression," *Journal of Social and Clinical Psychology* 37(10): 751–68. https://doi.org/10.1521/jscp.2018.37.10.751

Ibrahim, Sara. 2021. "How deepfakes are impacting our vision of reality," *Swissinfo*, August 13, 2021. https://www.swissinfo.ch/eng/how-deepfakes-are-impacting-our-vision-of-reality/46862004

Kay, Grace, and Isobel Asher Hamilton. 2022. "Google engineer believed chatbot had become an 8-year-old girl. Experts say it's not sentient – just programmed to sound 'real,'" *Business Insider*, June 13, 2022. https://www.businessinsider.com/lamda-ai-isnt-sentient-google-engineer-claims-2022-6

Kross, Ethan, Philippe Verduyn, Emre Demiralp, Jiyoung Park, David Seungjae Lee, Natalie Lin, Holly Shablack, John Jonides, and Oscar Ybarra. 2013. "Facebook use predicts declines in subjective well-being in young adults," *PLoS ONE* 8(8): e69841.

Kuta, Sarah. 2022. "Art Made With Artificial Intelligence Wins at State Fair," *Smithsonian Magazine*, September 6, 2022. https://www.smithsonianmag.com/smart-news/artificial-intelligence-art-wins-colorado-state-fair-180980703/

Lang, Rachel. 2023. "People are blown away by T-Pain's singing voice without auto-tune," *UNILAD*, last updated May 19, 2023. https://www.unilad.com/celebrity/people-are-blown-away-by-tpains-singing-voice-without-autotune-544092-20230519

Lee, Anthony. 2022. "AI or No, It's Always Too Soon to Sound the Death Knell of Art," *Wired*, December 27, 2022. https://www.wired.com/story/art-history-photography-painting-dalle-ai/

Leung, Wency. 2018. "How will AI technologies affect child development?" *The Globe and Mail*, July 22, 2018. https://www.theglobeandmail.com/life/article-how-will-ai-technologies-affect-child-development/

Liu, Chen-Chung, Mo-Gang Liao, Chia-Hui Chang, and Hung-Ming Lin. 2022. "An Analysis of Children's Interaction with an AI Chatbot and Its Impact on Their Interest in Reading," *Computers and Education* 189: 104576. https://doi.org/10.1016/j.compedu.2022.104576

Kingson, Jennifer A. 2022. "New AI tools let you chat with your dead relatives," *AXIOS*, July 13, 2022. https://www.axios.com/2022/07/13/artificial-intelligence-chatbots-dead-relatives-grandma

Madarang, Charisma, Kalia Richardson, and Krystie Lee Yandoli. 2023. "SAG-AFTRA Reveals How Studios Will Handle AI Replicas of Living and Dead Actors," *Rolling Stone*, November 10, 2023. https://www.rollingstone.com/tv-movies/tv-movie-news/sag-aftra-studio-deal-artificial-intelligence-actors-1234873708/

Madzarac, Milana. 2023. "Generative AI: ChatGPT enhances experiential learning in pharmacy," *University of Waterloo News*, February 27, 2023. https://uwaterloo.ca/news/science/generative-ai-chatgpt-enhances-experiential-learning

Marr, Bernard. 2023. "Picture Perfect: The Hidden Consequences of AI Beauty Filters," *Forbes*, June 06, 2023. https://www.forbes.com/sites/bernardmarr/2023/06/09/picture-perfect-the-hidden-consequences-of-ai-beauty-filters/?sh=147ee9d17d5d

Marsden, Rhodri. 2018. "How face filters on phone apps are leading teens to get plastic surgery," *The National*, August 15, 2018. https://www.thenationalnews.com/arts-culture/comment/how-face-filters-on-phone-apps-are-leading-teens-to-get-plastic-surgery-1.760364

Mercado, Melissa C., Kristin Holland, Ruth W. Leemis, Deborah M. Stone, and Jing Wang. 2017. "Trends in Emergency Department Visits for Nonfatal Self-inflicted Injuries Among Youth Aged 10 to 24 Years in the United States, 2001–2015," *JAMA* 318(19):1931–33. https://10.1001/jama.2017.13317

Metz, Rachel. 2022. "AI won an art contest, and artists are furious," *CNN Business*, September 3, 2022. https://www.cnn.com/2022/09/03/tech/ai-art-fair-winner-controversy

Mill, J.S. 1978. *On Liberty*. Indianapolis: Hackett Publishing.

Mills, Terance. 2021. "The Effects of AI on Child Psychology," *Forbes*, July 27, 2021. https://www.forbes.com/sites/forbestechcouncil/2021/07/27/the-effects-of-ai-on-child-psychology/?sh=6c6ffbc012f3

Miron, Oren, Kun-Hsing Yu, Rachel Wilf-Miron, and Isaac S. Kohane. 2019. "Suicide Rates Among Adolescents and Young Adults in the United States, 2000–2017," *JAMA* 321(23): 2362–64. https://10.1001/jama.2019.5054

Moore, Nick. 2023. "Artificial Intelligence voice scams on the rise, Better Business Bureau," *CTV News*, April 24, 2023. https://atlantic.ctvnews.ca/artificial-intelligence-voice-scams-on-the-rise-better-business-bureau-warns-1.6369696

Nussbaum, Martha. 2003. "Capabilities as Fundamental Entitlements: Sen and Social Justice," *Feminist Economics* 9(2–3): 33–59.

O'Neill, Natasha. 2023. "Can AI 'bring back' the dead? Debating the use of tech in the grieving process," *CTV News*, April 30, 2023. https://www.ctvnews.ca/sci-tech/can-ai-bring-back-the-dead-debating-the-use-of-tech-in-the-grieving-process-1.6370606

Parent, William A. 1983. "Privacy, Morality, and the Law," *Philosophy & Public Affairs* 12(4): 269–88.

Peters, Jay. 2023. "Microsoft says listing the Ottawa Food Bank as a tourist destination wasn't the result of 'unsupervised AI'," *The Verge*, August 17, 2023. https://www.theverge.com/2023/8/17/23836287/microsoft-ai-recommends-ottawa-food-bank-tourist-destination

Rajanala, Susruthi, Mayra B.C. Maymone, and Neelam A. Vashi. 2018. "Selfies – Living in the Era of Filtered

Photographs," *Jama Facial Plastic Surgery* 20(6): 443–44.

Raudsepp, L., and K. Kais. 2019. "Longitudinal associations between problematic social media use and depressive symptoms in adolescent girls," *Preventive Medicine Reports* 15:1–5.

Reynolds, Glenn Harlan. 2018. "Social media firms want us addicted to approval. So much for WiFi making us smarter," *USA Today*, April 1, 2018. https://www.usatoday.com/story/opinion/2018/04/01/social-media-business-model-addicts-us-approval-not-information-column/476719002/

Roberts, Millie. 2022. "The Euphoric Highs & Problematic Lows of AI Avatar Art," *Refinery 29*, last updated December 8, 2022. https://www.refinery29.com/en-au/ai-avatar-art

Romine, Taylor. 2022. "Police Play Disney tunes to prevent video of them on patrol being posted online, California lawmaker claims," *CTV News*, May 1, 2022. https://www.ctvnews.ca/world/police-play-disney-tunes-to-prevent-video-of-them-on-patrol-being-posted-online-california-lawmaker-claims-1.5884057

Roose, Kevin, 2022. "An A.I.-Generated Picture Won an Art Prize. Artists Aren't Happy," *The New York Times*, September 2, 2022. https://www.nytimes.com/2022/09/02/technology/ai-artificial-intelligence-artists.html

Schatten, Jeff. 2022. "Will Artificial Intelligence Kill College Writing?" *The Chronicle of Higher Education*, September 14, 2022. https://www.chronicle.com/article/will-artificial-intelligence-kill-college-writing

Schmunk, Rhianna. 2024. "Family of late comedian George Carlin sues podcast hosts over AI impression," *CBC News*, January 30, 2024. https://www.cbc.ca/news/canada/british-columbia/george-carlin-ai-podcast-lawsuit-1.7098925

Screen Time Action Network. 2018. "Open letter to Jessica Henderson, President of the American Psychological Association, August 8, 2018," *Screen Time Action Network*, accessed July 19, 2023. https://screentimenetwork.org/apa

Seighart, Mary Ann. 2022. *The Authority Gap*. London: Penguin Random House.

Shakya, Holly B., and Nicholas A. Christakis. 2017. "Association of Facebook Use With Compromised Well-Being: A Longitudinal Study," *American Journal of Epidemiology* 185(3): 203–11. https://doi.org/10.1093/aje/kww189

Sharf, Zach. 2023. "'Indiana Jones 5' De-Aged Harrison Ford With A.I. and Old Film Footage of Him That Lucasfilm Never Printed: 'That's My Actual Face,'" *Variety*, February 6, 2023. https://variety.com/2023/film/news/indiana-jones-5-artificial-intelligence-de-age-harrison-ford-unreleased-footage-1235514222/

Simonite, Tom. 2019. "How Deepfakes Scramble Our Sense of True and False," *WIRED*, December 7, 2019. https://www.wired.com/story/how-deepfakes-scramble-sense-true-false

Somos, Christy. 2021. "Alexa 'error' fixed after girl told to stick penny in socket for a 'challenge,'" *CTV News*, December 28, 2021. https://www.ctvnews.ca/sci-tech/alexa-error-fixed-after-girl-told-to-stick-penny-in-socket-for-a-challenge-1.5721080

Sulleyman, Aatif. 2017. "Facebook is 'destroying' society and making users feel 'vacant and empty,' former exec says," *Independent*, December 12, 2017. https://www.independent.co.uk/tech/facebook-society-destroy-social-media-network-users-damage-communicate-connect-chamath-palihapitiya-a8105131.html

Symons, John, and Ramón Alvarado. 2022. "Epistemic Injustice and Data Science

Technologies," *Synthese (Dordrecht)* 200(2). https://doi.org/10.1007/s11229-022-03631-z

Tiffany, Kaitlyn. 2022. "TikTok Has a Problem," *The Atlantic*, March 31, 2022. https://www.theatlantic.com/technology/archive/2022/03/west-elm-caleb-tiktok-mob-villain/629423/

Townsend, Chance. 2022. "James Earl Jones signs over rights to voice of Darth Vader to be replaced by AI," *Mashable*, September 24, 2022. https://mashable.com/article/james-earl-jones-gives-rights-to-darth-vader-ai

Twenge, Jean M., Gabrielle N. Martin, and W. Keith Campbell. 2018. "Decreases in Psychological Well-Being Among American Adolescents After 2012 and Links to Screen Time During the Rise of Smartphone Technology," *Emotion* (Washington, D.C.) 18(6): 765–80. https://doi.org/10.1037/emo0000403

Twenge, Jean M., Gabrielle N. Martin, and Brian H. Spitzberg. 2019. "Trends in U.S. Adolescents' Media Use, 1976–2016: The Rise of Digital Media, the Decline of TV, and the (Near) Demise of Print," *Psychology of Popular Media Culture* 8(4): 329–45. https://doi.org/10.1037/ppm0000203

Vincent, James. 2017. "Former Facebook exec says social media is ripping apart society," *The Verge*, December 11, 2017. https://www.theverge.com/2017/12/11/16761016/former-facebook-exec-ripping-apart-society

——. 2018. "Watch Jordan Peele use AI to make Barack Obama deliver PSA about fake news," *The Verge*, April 17, 2018. https://www.theverge.com/tldr/2018/4/17/17247334/ai-fake-news-video-barack-obama-jordan-peele-buzzfeed

——. 2021. "Tom Cruise deepfake creator says public shouldn't be worried about 'one-click' fakes," *The Verge*, March 5, 2021. https://www.theverge.com/2021/3/5/22314980/tom-cruise-deepfake-tiktok-videos-ai-impersonator-chris-ume-miles-fisher

Vind Jensen, Tore. 2019. "Abundance of information narrows our collective attention span," *DTU*, April 15, 2019. https://www.dtu.dk/english/news/all-news/nyhed?id=%7b246bbed3-8683-4012-a294-20db7f0015f4%7d

Vogels, Emily A., Risa Gelles-Watnick, and Navid Massarat. 2022. "Teens, Social Media and Technology 2022," *Pew Research Center*, August 10, 2022. https://www.pewresearch.org/internet/2022/08/10/teens-social-media-and-technology-2022/

Warren, Samuel D., and Louis D. Brandeis. 1890. "The Right to Privacy," *Harvard Law Review* 4(5): 193–220.

Webster, Andrew. 2023. "Actors say Hollywood studios want their AI replicas – for free, forever," *The Verge*, July 13, 2023. https://www.theverge.com/2023/7/13/23794224/sag-aftra-actors-strike-ai-image-rights

White, Jessica. 2022. "Inside the disturbing rise of 'deepfake' porn," *DAZED*, April 19, 2022. https://www.dazeddigital.com/science-tech/article/55926/1/inside-the-disturbing-rise-of-deepfake-porn

Wu, Tim. 2016. *The Attention Merchants: The Epic Scramble to Get Inside Our Heads*. New York: Alfred. A. Knopf.

Zhang, Baiwu, Jin Peng Zhou, Ilia Shumailov, and Nicolas Papernot. 2020. "On Attribution of Deepfakes." https://doi.org/10.48550/arxiv.2008.09194

Index

Page numbers in italics denote figures.

About the Publisher

The word "broadview" expresses a good deal of the philosophy behind our company. Our focus is very much on the humanities and social sciences—especially literature, writing, and philosophy—but within these fields we are open to a broad range of academic approaches and political viewpoints. We strive in particular to produce high-quality, pedagogically useful books for higher education classrooms—anthologies, editions, sourcebooks, surveys of particular academic fields and sub-fields, and also course texts for subjects such as composition, business communication, and critical thinking. We welcome the perspectives of authors from marginalized and underrepresented groups, and we have a strong commitment to the environment. We publish English-language works and translations from many parts of the world, and our books are available world-wide; we also publish a select list of titles with a specifically Canadian emphasis.

broadview press

This book is made of paper from well-managed FSC® - certified
forests, recycled materials, and other controlled sources.

FSC
www.fsc.org

MIX
Paper | Supporting
responsible forestry
FSC® C103567

PCF

BIO GAS®
ENERGY

∞
PERMANENT